A Primer of OILWELL DRILLING

A Basic Text of Oil and Gas Drilling

Seventh Edition

by Dr. Paul Bommer

in cooperation with

**INTERNATIONAL ASSOCIATION
OF DRILLING CONTRACTORS (IADC)**

published by

THE UNIVERSITY OF TEXAS

CONTINUING EDUCATION
PETROLEUM EXTENSION SERVICE

2008

Library of Congress Cataloging-in-Publication Data

Bommer, Paul–
    A primer of oilwell drilling. — 7th ed./ by Paul Bommer
       p. cm.
    ISBN 0-88698-227-8 (alk. paper)
    1. Oil well drilling. I. University of Texas at Austin.
Petroleum Extension Service. II. Title.
TN871.2B316 2001
622'.3382—dc21

                                         00-011902   CIP

Catalog No. 2.00070
ISBN 0-88698-227-8

*The University of Texas at Austin is an equal opportunity institu-
tion. No state tax funds were used to print this book.*

Contents

Figures   vi

Tables   xiv

Preface   xv

About the Author   xvii

1   **Introduction**   1

2   **History**   7
The Drake Well, 1850s   9
California, Late 1800s   10
The Lucas Well, 1901   11
The Middle East, 1900s   13

3   **Cable-Tool and Rotary Drilling**   15
Cable-Tool Drilling   15
Rotary Drilling   16
    Rotating Systems   18
    Fluid Circulation   19

4   **Rotary Rig Types**   21
Land Rigs   22
Mobile Offshore Rigs   23
    Bottom-Supported MODUs   24
    Floating Units   30

5   **People and Companies**   37
Operating Companies   38
Drilling Contractors   39
Drilling Contracts   39
Service and Supply Companies   40
People   44
    Drilling Crews   44
    Drilling Crew Work Shifts   50
    Crew Safety   50
    Other Rig Workers   51

6   **Oil and Gas: Characteristics and Occurrence**   55
Natural Gas   55
    Liquefied Natural Gas (LNG)   56
    Liquefied Petroleum Gas (LPG)   56
    Natural Gas Liquid (NGL)   57
Crude Oil   57
Refined Hydrocarbons   57
Oil and Gas Reservoirs   58
    Characteristics of Reservoir Rocks   58
    Origin and Accumulation of Oil and Gas   60
    Petroleum Traps   61
Types of Wells   69

7   **The Drill Site**   71
Choosing the Site   71
Preparing the Site   73

Surface Preparation   73
Earthen Pits   73
Cellars   77
Rathole   77
Mousehole   79
Conductor Hole   80
Moving Equipment to the Site   82
Moving Land Rigs   82
Moving and Setting Up Offshore Rigs   84

8 Rigging Up   85
Substructures   85
The Drawworks   88
Raising the Mast or Derrick   89
Derrick and Mast Heights   90
Mast Load Ratings   91
Rigging Up Additional Equipment   91
Offshore Rig-Up   92

9 Rig Components   93
Power System   93
Mechanical Power Transmission   97
Electrical Power Transmission   97
Hoisting System   100
The Drawworks   101
The Catheads   102
The Blocks and Drilling Line   104
Mast and Derricks   109
Rotating Systems   110
Rotary-Table System   110
Top Drive   117
Downhole Motors   118
The Drill String   120
Bits   122
Circulating System   126
Drilling Fluid   126
Circulating Equipment   128

10 Normal Drilling Operations   135
Drilling the Surface Hole   135
Tripping Out with a Kelly System   148
Tripping Out with a Top-Drive Unit   152
Tripping Out with a Pipe Racker   152
Running Surface Casing   154
Cementing   158
Tripping In   160
Drilling Ahead   162

11 Formation Evaluation   163
Examining Cuttings and Drilling Mud   163
Well Logging   165

Drill Stem Testing 168
Coring 170

12 Completing the Well 173
Plugging and Abandoning a Well 173
Completing a Producing Well 173
Production Tubing 174
Perforating 176
Well Testing and Treating 177
   Acidizing 177
   Fracturing 177
   Gravel Packing 178

13 Special Operations 179
Directional Drilling 179
   Slide Drilling with a Motor 180
   Rotary Steerable Assemblies 181
Fishing 181
Well Control 183

14 Rig Safety and Environmental Concerns 189

15 Conclusion 191

Appendix 1: Units of Conversion 193

Appendix 2: Figure Credits 195

Glossary 207

Index 233

Figures

1. Drilling rigs are large to accommodate the size of the drilling equipment and pipes.   1
2. ConocoPhillips Britannia platform in the North Sea   2
3. Drilling rig with a mast height of 147 feet (45 metres)   2
4. Personal protective equipment (PPE) includes hard hats, gloves, hearing protection, and safety glasses.   3
5. Steel stairways with handrails are used to get to the drilling rig floor. Note the drill pipe on the ramp at right.   4
6. The drawworks is part of the hoisting system used to lift drill pipe into place.   5
7. Whaling ships in New Bedford, Massachusetts. The barrels in the foreground are filled with whale oil.   7
8. Oilwells in Balakhani, a suburb of Baku, Azerbaijan, in the late 1800s   8
9. Oil Creek near Titusville, Pennsylvania as it looks today   8
10. Edwin L. Drake (right) and his good friend Peter Wilson, a Titusville pharmacist, in front of the historic Drake well in 1861   9
11. Patillo Higgins   11
12. Anthony Lucas, mining engineer at Spindletop   11
13. Wall cake stabilizes the drilling hole   12
14a. The 1901 Lucas well is estimated to have flowed about 2 million gallons (7,570 cubic metres) of oil per day.   12
14b. Spindletop oilfield in 1903, two years after the first well was drilled   12
15. A cable-tool rig   15
16. A polycrystalline diamond compact bit (PDC) (left) and a tri-cone bit (right)   17
17. The drill stem puts the bit on the bottom of the drilling hole.   17
18. Two floorhands place a joint of drill pipe in the mousehole prior to adding it to the active drill string.   17
19. Components in the rotary table rotate the drill string and bit.   18
20. A powerful motor in the top drive rotates the drill string and bit.   18
21. The bit is rotated by a downhole motor placed near it.   18
22. A pump circulates drilling mud down the drill pipe, out the bit, and up the hole.   19
23. Two pumps are available on this rig to move drilling fluid down the pipe.   19
24. Drilling mud   20
25. A land rig   21
26. An offshore jackup rig   21
27. An inland barge rig   21
28. Rigs can be disassembled and moved piece-by-piece to a new location.   23

29. Types of MODUs   23

30. The first MODU was a posted-barge submersible designed to drill in shallow water.   25

31. When the bottles are flooded, the weight makes the bottle-type rig sink to the seafloor.   25

32. Ice floes on the North Bering Sea   26

33. A concrete island drilling system (CIDS) features a reinforced concrete caisson.   26

34. Drilling equipment is placed on the deck of a barge to drill in the shallow waters of bays and estuaries.   27

35. Four boats tow a jackup rig to its drilling location.   28

36. A jackup rig with four column-type legs   28

37. A jackup with open-truss legs   29

38. The hulls of these jackups are raised to clear the highest anticipated waves.   29

39. A semisubmersible rig floats on pontoons.   30

40. The heavy lift vessel, Blue Marlin, transporting BP's semisubmersible, Thunder Horse   31

41. The pontoons of this semisubmersible float a few feet (metres) below the water's surface.   31

42. The main deck of a semisubmersible is huge. Shown here is the deck of the BP Thunder Horse.   32

43. Pathfinder 10,000-foot ultradeepwater drillship   33

44. Marine riser   34

45a. The heave compensator keeps proper tension on the drill string.   35

45b. Heave compensator   36

46. Workers on a drilling rig   37

47. U.S. Department of Interior Mineral Management Service map of proposed sale of government mineral leases in 2001   38

48. IADC standard drilling bid form   41

49. A computer display showing a well log   42

50. A member of a casing crew stabs one joint of casing into another.   43

51. Personnel on this offshore rig enjoy good food in the galley.   43

52. A driller on an offshore rig works in an environmentally controlled cabin.   45

53. The view from above the derrickman's position on the monkeyboard   46

54. A derrickman checking the weight or density of the drilling mud   47

55. Floorhands latch big wrenches called tongs onto the drill pipe.   48

56. Floorhands using power tongs to tighten drill pipe   49

57. Roustabouts move casing from a supply boat to the rig.   52

58. A crane operator manipulates controls from a position inside the crane cab.   53

59. A barge engineer monitors a semisubmersible's stability from a work station on board the rig.   53

60. BP's Thunder Horse listing in the Gulf of Mexico after a storm   54

61. Arctic Discoverer LNG transport ship   56

62. A pore is a small open space in a rock.   58

63. A cross-section showing pore space and the small connections between larger pores   58

64. Connected pores give rocks permeability.   59

65. A fault trap and an anticlinal trap   61

66. Types of stratigraphic traps   63

67. A combination trap   64

68. A piercement salt dome   64

69. To the right of the tire, a large heavy plate vibrates against the ground to create sound waves.   66

70. Several special trucks vibrate plates against the ground.   66

71. Fugro Explorer seismic vessel   67

72. Stuck into the ground, a geophone picks up reflected sound waves.   67

73. iZone Virtual Reality room at EPI Centre in Rijswijk, the Netherlands, 2008   68

74. Geologists working at a prospective petroleum area at the Peel Plateau in the Yukon   71

75. A reserve pit   74

76. Typical onshore layout of a drilling location   75

77a. Pit cleaning with Super Vac units   76

77b. Reserve pit cleanup and removal   76

78. A concrete pad to support the substructure surrounds this cellar.   77

79. The kelly has been placed in the rathole when the rig is not drilling.   78

80. A joint of drill pipe rests in this rig's mousehole.   79

81. A rathole rig drills the first part of the hole.   80

82. The conductor hole   80

83. The large diameter pipe to the right is the top of the conductor pipe.   81

84. A portable shallow oil drilling rig   83

85a. A heavy lift vessel carries a semisubmersible to a new drilling location.   84

85b. The Black Marlin heavy lift vessel transporting the Nautilus rig   84

86. A box-on-box substructure   86

87. A slingshot substructure is shown in folded position prior to being raised.   87

88. The slingshot substructure near its full height   87

89. This drawworks will be installed on the rig floor.   88

90. The drilling line is spooled onto the drawworks drum.   88

91. A mast being raised to a vertical position   89

92. This rig with a standard derrick was photographed in the 1970s at work in West Texas.   89

93. The derrick supports the weight of the drill string and allows the drill string to be raised and lowered.   90

94. The doghouse is located at the rig floor level.   91

95. In the foreground is a coal-fired boiler that made steam to power the cable-tool rig in the background.   93

96. A mechanical rig is shown drilling in West Texas in the 1960s.   94

97. Three diesel engines power this rig.   95

98. Three engines drive a chain-and-sprocket compound to power equipment.   96

99. The diesel engine at right directly drives an alternating current electric generator.   97

100. Controls in the SCR house where AC electricity is converted to the correct DC voltage for the many DC motors powering this rig.   98

101. A motor-driven drawworks   98

102. Two powerful electric DC traction motors drive the drawworks on this rig.   99

103. The hoisting system   100

104. The drawworks   101

105. Removing the drawworks housing reveals the main brake bands to the left and right on the hubs of the drawworks drum.   101

106. The electromagnetic brake is mounted on the end of the drawworks.   102

107. A floorhand has a fiber rope wrapped around a friction cathead to lift an object on the rig floor.   102

108. Floorhand using an air hoist to lift an object   103

109. This makeup cathead has a chain coming out of it that is connected to the tongs.   104

110. Wire-rope drilling line coming off the drawworks drum   105

111. Drilling line is stored on this supply reel at the rig.   105

112. Drilling line is firmly clamped to this deadline anchor.   105

113. The sheaves (pulleys) of this crown block are near the bottom of the photo.   106

114. Ten lines are strung between the traveling block and the crown block.   107

115. Several wraps of drilling line on the drawworks drum   107

116. Traveling block and kelly assembly   108

117. The mast supports the blocks and other drilling tools.   109

118.   A rotary-table system   110

119.   The turntable is housed in a steel case.   111

120.   The master bushing fits inside the turntable.   111

121.   Crewmembers are installing one of two halves that make
       up the tapered bowl.   112

122.   Crewmembers set slips around the drill pipe and inside the
       master bushing's tapered bowl to suspend the pipe.   113

123.   The master bushing has four drive holes into which steel
       pins fit on the kelly drive bushing.   113

124.   A master bushing with a square bottom that fits into a
       square opening in the master bushing   113

125a. A square kelly   114

125b. A hexagonal kelly   114

126.   A hexagonal kelly inside a matching opening in the top of
       the kelly drive bushing   114

127.   The hook on the bottom of the traveling block is about to
       be latched onto the bail of the swivel.   115

128   Drilling fluid goes through the rotary hose and enters the
       swivel through the gooseneck.   116

129.   A top drive, or power swivel, hangs from the traveling
       block and hook.   117

130.   Mud pressure pumped through the drill string forces the
       spiral rotor of the mud motor to turn inside the rubber
       helical-shaped stator.   118

131.   Horizontal hole   119

132.   A downhole motor lying on the rack prior to being run
       into the hole   119

133.   An adjustable bent housing on the motor deflects the bit
       a few degrees off-vertical to start the directional hole.   119

134.   Drill collars are placed on the pipe rack prior to being run
       in the hole.   120

135.   Drill collars put weight on the bit, which forces the bit
       cutters into the formation to drill it.   120

136.   Several joints of drill pipe are placed on the pipe rack
       before being run in the well.   121

137.   A floorhand stabs the pin of a joint of drill pipe into the
       box of another joint.   121

138.   Two drill collars on a pipe rack; at left is the drill collar box;
       at right is the pin   122

139.   Drill collars racked in front of drill pipe on the rig floor   122

140.   A roller cone bit has teeth (cutters) that roll, or turn, as the
       bit rotates.   123

141.   Tungsten carbide inserts are tightly pressed into holes
       drilled into the bit cones.   123

142.   Drilling fluid (salt water in this photo) is ejected out of the
       nozzles of a roller cone bit.   123

143.   Bit cutaway showing internal bearing   124

144.   Several types of natural diamond bits are available.   125

145.   Several diamond-coated tungsten carbide disks (compacts) form the cutters on this polycrystalline diamond compact (PDC) bit.   125

146.   Drilling mud swirls in one of several steel tanks on this rig.   126

147.   A derrickman measures the density (weight) of a drilling mud sample using a balance calibrated in pounds per gallon.   127

148.   Powerful mud pumps (most rigs have at least two) move drilling mud through the circulating system.   129

149.   Components of a rig circulating system   129

150.   The standpipe runs up one leg of the derrick, or mast, and conducts mud from the pump to the rotary hose.   130

151.   Mud with cuttings falls over the vibrating shale shaker screen.   131

152.   Desanders remove sand-sized particles from the mud.   131

153.   Desilters remove smaller silt-sized particles from the mud.   131

154.   The degasser removes a relatively small volume of gas that enters the mud from a downhole formation and is circulated to the surface in the annulus.   132

155.   A centrifuge removes particles even smaller than silt.   132

156.   A mud cleaner is used for mud weighted with barite.   133

157.   Bulk barite tanks with bagged chemicals in the foreground   133

158.   A derrickman, wearing personal protective equipment, adds dry components to the mud through a hopper.   134

159.   A closed-top chemical barrel for adding caustic chemicals to the mud in the tanks   134

160.   Typical wellbore architecture   135

161.   A bit being lowered into the hole on a drill collar   136

162.   A kelly with related equipment in the rathole   137

163.   Red-painted slips with three handgrips suspend the drill string in the hole.   137

164.   The kelly drive bushing is about to engage the master bushing on the rotary table.   138

165.   The motor in the top drive turns the drill stem and the bit.   138

166.   The black inner needle on the weight indicator shows the weight suspended from the derrick in thousands of pounds.   139

167.   The kelly is drilled down (close to the kelly drive bushing), and it is time to make a connection.   139

168.   Using the traveling block, the driller raises the kelly, exposing the first joint of drill pipe in the opening of the rotary table.   140

169.   Crewmembers latch tongs on the kelly and on the drill pipe.   141

170.   The kelly spinner rapidly rotates the kelly in or out of the drill pipe joint.   142

171.   Crewmembers stab the kelly into the joint of pipe in the mousehole.   143

172.   Crewmembers use tongs to buck up (tighten) one drill pipe joint to another.   144

173.   Crewmembers remove the slips.   145

174.   The kelly drive bushing is about to engage the master bushing.   145

175.   Making a connection with a kelly   145

176.   Making a connection using a top drive   146

177.   An Iron Roughneck™ spins and bucks up joints with built-in equipment.   147

178.   The kelly and swivel with its bail are put into the rathole.   148

179.   Crewmembers latch elevators to the drill pipe tool joint suspended in the rotary table.   149

180.   The floorhands set the lower end of the stand of pipe off to one side of the rig floor.   150

181.   The derrickman places the upper end of a stand of drill pipe between the fingers of the fingerboard.   151

182.   Making a trip   152

183.   Top view of an automatic pipe handling device manipulating a stand of drill pipe   153

184.   A casing crewmember cleans and inspects the casing as it lies on the rack next to the rig.   154

185.   Casing threads have been cleaned and inspected.   154

186.   A joint of casing being lifted onto the rig floor   155

187.   A joint of casing suspended in the mast; note the centralizer   155

188.   Casing elevators suspend the casing joint as the driller lowers the joint into the casing slips.   155

189.   Working from a platform called the stabbing board, a casing crewmember guides the casing elevators near the top of the casing joint.   156

190.   Crewmembers lift the heavy steel-and-concrete guide shoe.   157

191.   The guide shoe is made up on the bottom of the first joint of casing to go into the hole.   157

192.   Cementing the casing: (A) the job in progress; (B) the finished job   157

193.   Crewmembers install a float collar into a casing string.   157

194.   Scratchers and centralizers are installed at various points in the casing string.   158

195.   Top view of casing that is not centered in the hole.   158

196. A cementing head (plug container) rests on the rig floor, ready to be made up on the last joint of casing to go into the hole.   159

197. To trip in, crewmembers stab a stand of drill pipe into another.   160

198. After stabbing the joint, crewmembers use a spinning wrench to thread the joints together.   161

199. After spin up, crewmembers use tongs to buck up the tool joints to the correct torque.   161

200. A handful of cuttings made by the bit   163

201. Mud log section showing a formation that contains hydrocarbons   165

202. Logging personnel run and control logging tools by means of wireline from a logging unit.   166

203. A well-site log is interpreted to give information about the formations drilled.   167

204. Drill stem test tools   168

205. A successful DST   169

206. Repeat formation tester (RFT) tool   169

207a. A whole core barrel   170

207b. Sidewall coring device   170

208. A. An oil-saturated whole core from a South Texas well; B. Sidewall cores   171

209. This collection of valves and fittings is a Christmas tree.   173

210. Subsea wellheads   174

211. A coiled-tubing unit runs tubing into the well using a large reel.   175

212. Perforations (holes)   176

213. Shaped charges in a perforating gun make perforations.   176

214. A gravel pack   178

215. Several directional wells tap an offshore reservoir.   179

216. An overshot   181

217. A. The spear goes inside the fish in a released position. B. Once in position, the spear is set and the fish is removed.   182

218. Fluids erupting from underground caught fire and melted this rig.   183

219. A stack of BOPs installed on top of the well   185

220. Ram cutaway   185

221. A subsea stack of BOPs being lowered to the seafloor from a floating rig   186

222. Several valves and fittings make up a typical choke manifold.   186

223. A remote-controlled choke installed in the choke manifold   187

224. This control panel allows an operator to adjust the size of the choke.   187

Tables

1. Land Rigs Classified by Drilling Depth   22
2. Types of MODUs   24
3. IADC Annual Work Time and Accident Statistics   189

Preface

The Petroleum Extension Service (PETEX) published the first edition of *A Primer of Oilwell Drilling* in 1951. With this latest printing there have been seven editions of the primer written by several editors and authors. Each edition was created in order to keep the book current with advances in drilling technology.

Although drilling technology continues to evolve the purpose of this book has remained the same: to clearly explain drilling to non-technical readers. The book also includes sections on the history of the petroleum industry as well as the evolution of the science and art of drilling. Anyone with an interest in the oil and gas business in general and drilling in particular will find this a useful first reader on the subject. Additional information on the petroleum industry can be found in many of the other excellent books offered by PETEX.

This edition is a major revision of the works that came before. The task was made infinitely easier because of the excellent frame work built into the sixth edition by Ron Baker (then the Director of PETEX).

The manuscript was created certainly not just through me and my predecessors but by the excellent and supportive staff of PETEX. In particular I wish to thank Dr. Larry Lake, Chairman of my Department, for suggesting I become involved in this project and Ms. Francisca Kennedy-Ellis, Assistant Director of PETEX, who agreed.

PETEX is solely responsible for the contents of this book. While every effort has been made to ensure the accuracy of the contents, the book is intended only as a training aid and does not intend to approve or disapprove any specific product, service, or practice.

Paul M. Bommer, Ph.D.
Senior Lecturer
The Department of Petroleum
  and Geosystems Engineering
The University of Texas at Austin
2008

About the Author

Paul M. Bommer is a Senior Lecturer in Petroleum Engineering at The University of Texas at Austin. He received his Bachelor's ('76), Master's ('77), and Doctoral ('79) degrees in Petroleum Engineering, all from The University of Texas at Austin.

He spent over twenty-five years in industry as an oil and gas operator and consultant in Texas and other parts of the United States. He and his brother Peter (UT, BS-PGE '78) are co-owners in the firm of Bommer Engineering Company.

He is a third generation oil man following his father (UT, BS-PGE, '50) who was a highly regarded petroleum engineer in Texas as the principal owner of Viking Drilling Company in San Antonio and his paternal grandfather who was a field superintendent in Oklahoma, East Texas and on the Texas Gulf Coast for Stanolind (later Amoco) Oil Company. As with most oilfield families, his mother (UT, BS-HEc, '49) made sandwiches for the crews, curtains for the tool pusher's trailer, created a home, and raised the kids.

Introduction

This book is an introduction to the art and science of drilling oilwells. While this book focuses on well drilling in the *oil* and gas industry, it is important to note that *wells* can be drilled for a variety of purposes. Not all wells are used to extract oil and gas from the earth. Wells are also drilled to produce fresh water for irrigation and to supply water to cities. Some wells are drilled into deep layers of rock to dispose of hazardous waste. Greenhouse gases, such as carbon dioxide, can be captured and injected into underground layers for permanent disposal. The same well drilling methods can be applied to all these uses.

Drilling *rigs* are large and noisy. They operate numerous pieces of enormous equipment (fig. 1). The purpose of a drilling rig is only to drill a hole in the ground. Although the rig is big, the hole it drills is relatively small. The purpose of the drill hole is to *tap* an oil or gas *reservoir* often thousands of feet or hundreds of metres below the surface of the earth. The drill hole is usually less than one foot (30 centimetres) in diameter at final depth.

Figure 1. Drilling rigs are large to accommodate the size of the drilling equipment and pipes.

Drilling rigs can operate on land and *offshore*, which is the common oilfield term for drilling rigs at sea. *Offshore rigs* can be located many miles (kilometres) from shore and require transport by boat or helicopter to reach them (fig. 2).

The most distinctive characteristic of a drilling rig is the tall, strong, structural tower called a *mast* or *derrick* (fig. 3). Derricks must support the tremendous weight of the drilling tools, which can weigh many *tons (tonnes)*. Rig masts and derricks are built tall to accommodate the long lengths of drill pipe used in the drilling process. A mast or derrick can be as high as a 16-story building— about 200 feet (60 metres) tall. Generally, most derricks or masts are approximately 140 feet (43 metres) high.

COURTESY OF CONOCOPHILLIPS

Figure 2.   ConocoPhillips Britannia platform in the North Sea

Figure 3.   Drilling rig with a mast height of 147 feet (45 metres)

Personnel on drilling rigs have job titles unique to the industry. *Toolpusher* is the traditional term for the rig boss. The term describes the person who supervises a crew of workers and makes sure work is done on time. In the past, the toolpusher was required to push the crew to accomplish the work. Because drilling crews use a variety of tools to drill, the term toolpusher was adopted. The drilling industry now prefers to use the title *rig superintendent* or *rig manager* for the person in charge. However, on the drilling rig, the *rig hands* will still call him or her the toolpusher or, in Canada, the *toolpush*. Rig hands are the workers who provide the manual power necessary to carry out the work.

The rig superintendent informs all new hires and visitors about the rig's current activities and safety rules. Whether working on a rig or visiting, everyone must wear *personal protective equipment (PPE)* (fig. 4). Rig workers must wear gloves to protect their hands. Some companies require special fire-resistant clothing or coveralls. These special coveralls also have reflective strips on the arms, legs, and back so personnel can be seen after dark. (See photos of workers wearing PPE in Chapter 3. Cable-Tool and Rotary Drilling). A hard hat, steel-toed boots, safety glasses, and ear plugs are also part of the mandatory PPE on the job.

Figure 4. Personal protective equipment (PPE) includes hard hats, gloves, hearing protection, and safety glasses.

The *rig floor* is the main work area of the rig. It usually rests on a strong foundation or *substructure* that raises it above the ground or sea level. On the rig floor, the crews handle the lengths, or *joints*, of *drill pipe* that put the *bit* or the hole-boring device on the bottom of the hole. A rig floor might be as high as 40 feet (12 metres) above the ground or 100 feet (30 metres) above sea level. Steel stairs lead to the rig floor (fig. 5). Handrails ensure worker safety because these exterior stairs can be slick.

Figure 5. Steel stairways with handrails are used to get to the drilling rig floor. Note the drill pipe on the ramp at right.

When the rig is drilling, or *making hole*, there are distinctive loud squawks or squeals from the *drawworks brake* as it releases more drilling line, or *slacks off*, to allow the bit to *drill ahead*. The drawworks is a powerful *hoist* that regulates the weight the *drill string* puts on the bit (fig. 6). Every time the friction *brake bands* ease the grip on the steel hubs of the *drawworks drum*, there is a loud screech. Then, more drilling line is released from the *drum*, which lowers the drill pipe or drill string. By lowering the drill pipe, some of the drill string weight is applied to the bit at the bottom of the hole. Although a noisy process, it usually signals the bit is drilling ahead without problems.

Figure 6.   The drawworks is part of the hoisting system used to lift drill pipe into place.

What happens on the drilling rig is vital to the oil and gas industry. There are many operations besides drilling involved in getting *crude oil* and *natural gas* out of the ground and turned into marketable products such as gasoline and heating fuel. However, without a simple drilled hole in the ground, oil companies could not obtain petroleum at all.

This book identifies the people and tools used in the drilling process and provides an overview of oilwell drilling. To understand drilling today, a look back at the history of oilwell drilling is helpful.

The story of modern oilwell drilling began at the start of the industrial revolution. Workers wanted better ways to illuminate their homes when they returned from the factories. The steam-powered industrial machines increasingly used in factories also required good quality lubricant oils.

Responding to the demand for reliable lighting, companies began making oil lamps, which were brighter than candles, lasted longer, and were not easily blown out by errant breezes. The best source of oil to burn in the early oil lamps was sperm whale oil. Whale oil was clear, almost odorless, light in weight, and burned with little smoke.

While everyone preferred whale oil, by the mid-1800s it was so scarce that only the wealthy could afford it (fig. 7). Whalers in the New England region of the United States had nearly hunted sperm whales into extinction. There was a demand for something to replace whale oil.

Oil seeping out of shallow accumulations is a common, worldwide phenomenon. The area around Baku, Azerbaijan, had been known from ancient times to hold oil and natural *gas seeps*. The first modern oilwell was drilled in Baku in 1846. This well was drilled to a depth of 69 feet (21 metres). By 1872, due mainly to lamp oil demand, the Baku area had so many wells that it became known as the "Black City."

2
History

Source: NOAA, Dept. of Commerce

Figure 7. Whaling ships in New Bedford, Massachusetts. The barrels in the foreground are filled with whale oil.

COURTESY OF BRITA ÅSBRINK COLLECTION

Figure 8.   Oilwells in Balakhani, a suburb of Baku, Azerbaijan, in the late 1800s

By the beginning of the 20th century, Baku was responsible for half the world's oil produced until that time (fig. 8).

The quest for a better source of lubricants and lighting was also important in the United States. In 1854, George Bissell, a New York attorney, received a sample of an unusual liquid from a chemistry professor at Dartmouth College in New Hampshire. Bissell and the chemistry professor were interested in finding a whale oil substitute. The professor wanted Bissell's opinion of the sample liquid's value as a lamp oil and lubricant.

The sample had been collected near a creek that flowed through the woods on the Hibbard Farm in Venango County in northwestern Pennsylvania. The creek water carried an odorous, dark-colored substance that sometimes caught fire. When applied to machinery, the dark substance was a good lubricant. Because the substance flowed out of the nearby rocky land and in the creek, people called it *rock oil*. So much oil flowed into the stream that Pennsylvania settlers named it Oil Creek (fig. 9). The professor's sample came from the land next to the creek southeast of the town of Titusville, where the oil seeped from rocks in the water.

Figure 9.   Oil Creek near Titusville, Pennsylvania, as it looks today

THE DRAKE WELL, 1850s

After examining the oil sample, Bissell was convinced that refined rock oil would burn as cleanly and safely as any of the oils available at the time, including whale oil. He also believed that it would be a marketable lubricant. Bissell, James M. Townsend, a Connecticut banker, and several others formed what became the Seneca Oil Company in New Haven, Connecticut.

A problem the company faced was how to best extract the oil from the land. It was not efficient to simply wait for the oil to flow out of the rocks and skim it from the surface of the creek. Those who had collected oil using this method could only scoop up a gallon (a few litres) or two a day. Seneca Oil's goal was to produce large amounts of oil and market it in the populous northeastern United States.

Someone at Seneca Oil—no one knows exactly who—came up with the idea of drilling a well to tap the oil. Drilling was not a new concept. People had been drilling saltwater wells in the Titusville area for years. The salt water was dehydrated to produce salt and, in the time before refrigeration, the *brine* was also used as a food preservative. Many of the saltwater wells also produced oil, which the salt drillers considered a nuisance because the oil contaminated the salt.

James Townsend hired Colonel Edwin L. Drake to drill a well for Seneca Oil at the Oil Creek site. Drake, an ex-railroad conductor, used the honorary title of Colonel bestowed on him by Townsend. In the spring of 1859, Drake employed William A. Smith to be his well *driller*. Smith was a blacksmith and an experienced brine well driller. He showed up at the *well site* in Titusville with his sons as helpers, and his daughter served as the camp cook. One of the first things Drake and Smith did was to drive a length of hollow steel *casing pipe* through the soft surface soil until it reached *bedrock*. The casing prevented loose topsoil from *caving in* to the hole. To this day, drillers still begin wells by casing the top of the hole. Drake and Smith then built the drilling rig, ran the drilling tools inside the casing, and drilled the rock (fig. 10).

Drake and Smith had drilled the hole to a depth of about 69 feet (21 metres) when Smith noted that the bit had suddenly dropped 6 inches (15 centimetres). It was near quitting time on a Saturday, so he shut the operation down, assuming that he and the boys would continue drilling on the following Monday.

Figure 10. Edwin L. Drake (right) and his good friend Peter Wilson, a Titusville pharmacist, in front of the historic Drake well in 1861

Source: Pennsylvania Department of Conservation and Natural Resources

Sunday, August 27, 1859, was the driller's day off, but Smith decided to check on the well. He looked into the top of the casing and found the hole full of oil. Overnight, oil from a *formation* 69½ feet (21.2 metres) below the surface had flowed into the well casing and filled it to the top, indicating the drilling was a success. It is not known how much oil the well produced, but it was probably around 300 to 400 gallons (about 1,135 to 1,514 litres) per day. This was far more than the gallon or two that could be collected out of the seeps in the creek. The success of Drake's well demonstrated that a drilled well in the United States could yield marketable amounts of oil.

Drake's was the first well in the United States drilled for the sole purpose of finding and producing oil. News of the accomplishment spread rapidly and, because a ready market existed for refined rock oil, dozens of new rigs sprang up in the area to take advantage of demand. Saltwater drillers, formerly reluctant to drill oilwells, quickly changed their minds.

Colonel Drake's well in Titusville marked the beginning of the *petroleum* era in the United States. Refined rock oil became the primary lamp oil in homes and businesses. As industrial machines became more common, refined rock oil also became an important commercial lubricant.

CALIFORNIA, LATE 1800s

Reports of oil drilling in Pennsylvania rapidly reached all parts of the United Sates, Canada, and abroad. Interest in oilwell drilling was particularly high in California, where the population was rapidly growing. After prospectors found gold at Sutter's Mill in 1849, immigrants flooded into California. Unlike the northeastern United States, which had plenty of coal for heating and for firing industrial boilers and machinery, California had none. Fortunately, California had many oil and gas seeps similar to those in Pennsylvania.

Enterprising Californians applied drilling technology to the oil seeps in California. The first successful well was drilled in 1866 at Rancho Ojai near Ventura, California, and the Sulphur Mountain oil seeps. It was 550 feet (168 metres) deep and produced 15 to 20 barrels (about 2 to 3 cubic metres) a day. It was considered a great success and prompted the drilling of many more wells. Oil and gas *production* provided much of California's energy.

THE LUCAS WELL, 1901

The United States grew increasingly dependent on oil as a plentiful and inexpensive source of energy. Individuals and companies were drilling wells all over the country, and Texas was no exception. Beaumont, Texas, is located in flat, coastal-plain country. In the late nineteenth-century, Big Hill, whose formal name was *Spindletop*, was a dome rising about 15 feet (4.5 metres) above the surrounding plain. Gas seeping out of the dome could be ignited easily with a lighted match.

One person particularly interested in Spindletop was Patillo Higgins, a real estate speculator and self-taught geologist who lived in the area. He was convinced that oil and gas lay about 1,000 feet (305 metres) below Spindletop. Around 1890, Higgins purchased land on top of the Big Hill dome and, with several financial partners, drilled two unsuccessful wells (fig. 11).

The unsuccessful wells encountered a problem at about 350 feet (106 metres), when the bit hit a 250-foot- (76.2-metre) thick sand formation the drillers called *running quicksand*. The sand was so loose it caved into the drilled hole, making further drilling impossible. *Cable-tool drillers* ran casing, just as Drake had, attempting to prevent the cave-in. The formation was so thick it collapsed before it could be drilled or *cased* using cable-tool technology. Discouraged but certain that oil lay below Spindletop, Higgins offered to lease the property to anyone willing to drill a 1,000-foot (305-metre) well.

An Austrian mining engineer, named Anthony Lucas answered Higgins's call (fig. 12). Lucas visited Spindletop and agreed with Higgins that the hill was a salt dome surrounded by geologic formations that trapped oil and gas. After another frustrating and costly failure, Lucas finally *spudded* (began drilling) a new well at Spindletop on October 27, 1900. He hired the Hamil Brothers of Corsicana, Texas, to drill the well. The Hamil's equipment was a *rotary drilling rig* while most drillers used *cable-tool rigs*. Unlike cable-tool rigs, *rotary rigs* require *drilling fluid* to operate, and particles in the drilling fluid prevent formations from caving. Aware that the running quicksand would cause trouble, the Hamils paid close attention to the mix of their drilling fluid.

On the rotary rig used by the Hamils, drilling fluid was sent down the hole and picked up the rock *cuttings* made by the bit, then the fluid carried them back to the surface for disposal. The early rigs Drake and the early California drillers used did not require drilling fluid, dooming such rigs to extinction.

At Spindletop, the Hamils used water as a drilling fluid. The Hamils knew from their earlier drilling experience that clear water alone would not do the job. They hand dug a pit in the ground next to the rig, filled it with water, and pumped the water into the well as they drilled it. They needed to add *mud* because the tiny solid particles of *clay* in the muddy water would increase the *density* of the liquid.

Figure 11. Patillo Higgins

Figure 12. Anthony Lucas, mining engineer at Spindletop

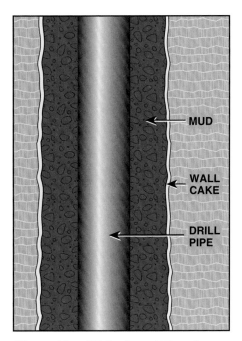

Figure 13.   Wall cake stabilizes the drilling hole.

Figure 14a.   The 1901 Lucas well is estimated to have flowed about 2 million gallons (7,570 cubic metres) of oil per day.

The resulting density of the mud was higher than that of water and this difference caused the particles in the clay mud to stick to the sides of the hole. The mud particles formed a thin, but strong sheath—a *wall cake*—on the sides of the hole, much like plaster on the walls of a room. Because the well was kept full of drilling fluid and the fluid formed a wall cake, the sand was stabilized, preventing it from caving in (fig. 13).

By January 1901, the new well reached about 1,000 feet (305 metres). On January 10, the crew began lowering a new bit to the bottom of the hole. Suddenly, *drilling mud* spewed out of the well. A geyser of oil soon followed it. It gushed 200 feet (60 metres) above the 60-foot-high (18-metre-high) derrick (figs. 14a and 14b).

As Lucas watched the *gusher* from a safe distance, he estimated it flowed at least 2 million gallons (nearly 7,570 cubic metres) of oil per day. In oilfield terms, that is about 48,000 barrels of oil per day. One *barrel* of oil is equal to 42 U.S. gallons.

Spindletop flowed unparalleled amounts of oil. It showed the effectiveness of a rotary-type rig, which was previously used infrequently by drillers. The Lucas well demonstrated that rotary rigs could drill wells that cable-tool rigs were incapable of drilling. Consequently, oilwell drillers began using rotary rigs more frequently than cable-tool rigs.

Figure 14b.   Spindletop oilfield in 1903, two years after the first well was drilled

THE MIDDLE EAST, 1900s

Many nations outside the United States were also using oil for fuel, lamp oil, and lubricants. World industrial production relied mainly on coal until fuel oil became commonly used. Fuel oil had several advantages over coal. The major advantage was that fuel oil had a higher energy content (*Btu*) than coal and was easier to ship and store. In 1911, the British began powering the ships of the Royal Navy with fuel oil rather than coal, making refueling at sea possible. The United States followed the British lead in converting its own Navy to using fuel oil. The use of fuel oil greatly reduced the time and manpower needed to refuel, supplied a more powerful fuel, and reduced the expense and problems of providing coaling stations around the world.

The United States had plentiful supplies of oil while the British did not. The British encouraged the formation of the Anglo-Persian Oil Company, which was the forerunner of British Petroleum. After exploring Persia (now Iran) and making agreements with the Persians, the Anglo-Persian Oil Company discovered a large oilfield with a well at Masjed Soleiman, Iran, in 1908. This was the first major oil discovery in the Middle East. Soon the Anglo-Persian Oil Company was the largest oil producing company in the world. Just as the Lucas well had demonstrated in the United States, the great Middle East oilfields showed that large volumes of oil could be developed on a global scale.

Cable-Tool and Rotary Drilling

Cable-tool drilling and rotary drilling techniques have been available since people first began making holes in the ground. Rotary rigs dominate the industry today, but cable-tool rigs drilled many wells in the past. Over 1,600 years ago, the Chinese drilled wells with various primitive yet efficient cable-tool rigs, which they continued to use into the 1940s. To quarry rocks for the pyramids, the ancient Egyptians drilled holes using hand-powered rotating bits. They drilled several holes in a line and stuck dry wooden pegs in the holes. Then they saturated the pegs with water. The swelling wood split the stone along the line made by the holes.

Most wells today are drilled with rotary rigs based on the Hamil Brothers' design at Spindletop.

CABLE-TOOL DRILLING

A steam-powered cable-tool rig was used by Drake and Smith to drill the Oil Creek site in Pennsylvania. The early drillers in California and elsewhere also used cable-tool rigs. The principle of cable-tool drilling is the same as that of a child's seesaw. When a child is on each end of a seesaw, it moves it up and down. The rocking motion demonstrates the principle of cable-tool drilling.

To explore the concept further, one could tie a cable to the end of the seesaw and let the cable dangle straight down to the ground. Next, a heavy chisel with a sharp point could be attached to the dangling end of the cable. By adjusting the cable's length so the end of the seesaw is all the way up, the chisel point hangs a short distance above the ground. Releasing the seesaw lets the heavy chisel hit hard enough to punch a hole in the ground. Repeating the process and rocking the seesaw causes the chisel to drill a hole. The process is quite effective. A heavy, sharp-pointed chisel can slowly force its way through rock, bit by bit, with every blow (fig. 15).

A cable-tool rig operates much like a seesaw with a powered *walking beam* mounted on a derrick. The walking beam is a wooden bar that rocks up and down on a central pivot. At Drake's rig, a *6-horsepower* (4.5-kilowatt) steamboat engine powered the walking beam. As the beam rocks up, it raises the cable attached to a chisel, or bit. Then, when the walking beam rocks down, heavy weights above the bit, called *sinker bars*, provide weight to ram it into the ground. The bit punches its way into the rock, and repeated lifting and dropping make the bit drill into the earth. The driller lets out the cable gradually as the hole deepens. The derrick provides space to raise the cable and pull the long drilling tools out of the hole using one of several winches called the *bullwheel*.

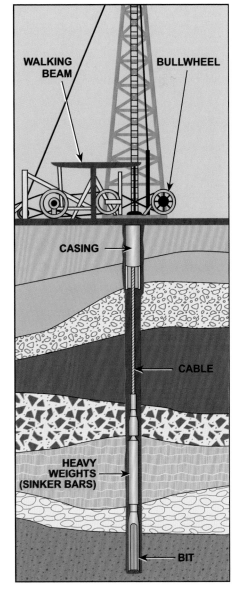

Figure 15.   A cable-tool rig

Despite cable-tool drilling's widespread use in the early days of *oilwell* drilling, the system had drawbacks. A major problem was that cable-tool drillers had to periodically stop drilling and remove the pulverized rock pieces from the bottom of the well. The bit had to be pulled out of the hole and a special basket, called a *bailer*, would be run into the hole. The bailer retrieved and removed the rock cuttings made by the bit. After bailing the cuttings, the bit was run back to the bottom to resume drilling. If the crew failed to bail out enough cuttings, the cuttings continued to obstruct the bit's progress. Bailing cuttings was not a significant hindrance because the wells were not deep and the cable-tool system allowed the crew to do the work fairly fast. The bailer cable was wound onto a winch, called the *sand reel*, and the crew simply reeled cable on and off to raise and lower the bailer. Reeling cable was a quick operation.

Cable-tool drilling worked well in the hard-rock formations like those in the eastern United States, the Midwest, and California. A big problem was that the cable-tool technique did not work in soft formations like clay or loose sand. Clay and sand closed around the bit and wedged the bit in the hole. Soft parts of the well that had not been cased often caved in, partially filling the hole. This limitation led to the increased use of rotary rigs, because more wells were being drilled in geographic locations like Spindletop where cable-tool bits got stuck. The drilling fluid used in rotary rigs prevented collapse in clay or sand formations.

The use of cable-tool rigs peaked in the 1920s and faded thereafter. One remaining use for cable-tool rigs today is to begin, or spud, the hole before the large rotary rig arrives. In this application, the cable-tool rig is called a spudder.

ROTARY DRILLING

Rotary drilling is different from cable-tool drilling. A rotary rig uses a bit that is completely different from a cable-tool chisel bit. A rotary bit has rows of teeth or other types of cutting devices that penetrate the formation and then scrape or gouge out pieces of it as the rig rotates the bit (fig. 16).

In addition, a rotary rig does not use a cable to suspend the bit in the hole. Rotary crewmembers attach the bit to the end of a long string of hollow pipe called a drill string. By screwing together several joints of pipe, the bit can reach the bottom of the hole (fig. 17). As the hole deepens, crewmembers add joints of pipe (fig. 18).

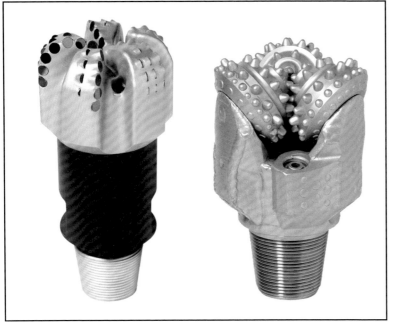

Figure 16.   *A polycrystalline diamond compact bit (PDC) (left) and a tri-cone bit (right)*

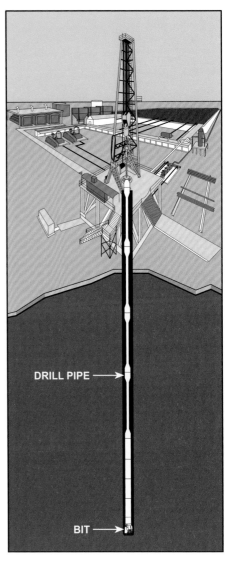

Figure 17.   *The drill stem puts the bit on the bottom of the drilling hole.*

Figure 18.   *Two floorhands place a joint of drill pipe in the mousehole prior to adding it to the active drill string.*

Figure 19. Components in the rotary table rotate the drill string and bit.

Rotating Systems

With the bit on the bottom of the hole, the rig can rotate the bit by using one of three different equipment systems. Many rigs use equipment called a *rotary table*, which is a kind of heavy-duty *turntable* (fig. 19).

Other rigs rotate the bit with a *top drive*, which is a piece of equipment with a powerful built-in electric or *hydraulic* motor that turns the pipe and bit (fig. 20).

In a few cases, a narrow downhole motor powered by the pressure of the drilling fluid rotates the bit. The motor is held by a long metal housing with a diameter a little less than that of the bit. The bit screws onto the end of it. A downhole motor can be used alone or along with either a rotary table or a top drive (fig. 21).

Today, most large rotary rigs use a top drive to rotate the pipe and bit. However, rigs using rotary tables have been around for a long time. Many drilling companies still own and use rotary tables because they are simple, rugged, and easy to maintain. Rotary rig owners often use downhole motors when they want to rotate the bit without rotating the entire string of pipe, or when they want to increase the speed available from the surface equipment.

Figure 20. A powerful motor in the top drive rotates the drill string and bit.

Figure 21. The bit is rotated by a downhole motor placed near it.

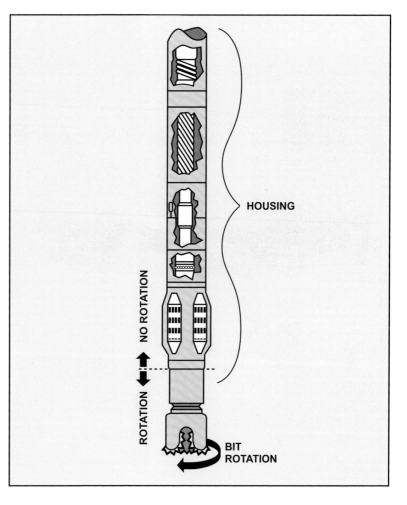

Regardless of the system used to rotate the bit, the driller allows some of the weight of the pipe to press down on the bit. The weight causes the bit cutters to bite into the formation rock. Then, as the bit rotates, the cutters roll over the rock and scrape or gouge it out.

Fluid Circulation

Just rotating the bit at the end of a pipe will not get the drilling job done by itself. The cuttings made by the bit must be removed. Otherwise, the cuttings collect under the bit and obstruct the drilling process. On a cable-tool rig the crews stop drilling to bail the cuttings from the hole. However, on a rotary rig, the crew does not have to stop drilling to bail cuttings because the rotary rig *circulates* fluid while the bit drills. The drilling fluid carries the cuttings up to the surface.

Crewmembers attach a rotary bit to hollow pipe instead of to braided cable. The hollow pipe is a channel for the drilling fluid. A powerful *pump* on the surface moves fluid down the pipe to the bit and back to the surface (fig. 22).

The circulating drilling fluid picks up the cuttings as the bit produces them. The cuttings are carried to the surface where they are separated from the fluid. The pump then moves the clean fluid back down the hole to repeat the cycle (fig. 23).

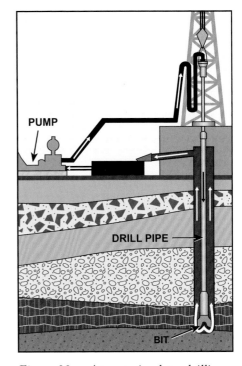

Figure 22.   A pump circulates drilling mud down the drill pipe, out the bit, and up the hole.

Figure 23.   Two pumps are available on this rig to move drilling fluid down the pipe. Normally only one at a time is used. If more volume is needed, both pumps are used.

The drilling fluid is usually a special liquid called drilling mud (fig. 24). The earliest drilling muds were plain, watery mud. Legend has it that at Spindletop, the Hamils created drilling mud by running cattle through the earthen pit to stir up the clay and muddy the water.

Drillers prepare, or *condition*, the drilling mud to control the specific formations into which they are drilling. Modern drilling mud is a much more complex blend of materials. Sometimes drilling mud is not even liquid. A better name for drilling mud is drilling fluid. Drilling fluids can actually be a liquid, gas, or foam.

The main advantage of a rotary rig is that crews do not have to be concerned about soft formations caving in on the bit and blocking the drill. Besides keeping *boreholes* from caving in, circulating drilling fluid performs several other important functions. For example, it moves the cuttings away from the drilling bit and cools and lubricates the bit. It also keeps *formation fluid*s from entering the hole and blowing out to the surface. There are several kinds of rotary rigs available for drilling on land and offshore.

Figure 24.   Drilling mud

A variety rotary drilling rigs might be used depending on the location and geography of the reservoir.

Offshore, the ocean environment plays an important role in rig design. Rigs may be broadly divided into two categories: rigs that work on land (fig. 25) and rigs that work offshore (figs. 26 and 27).

One type of offshore drilling facility is a *platform*. Although drilling occurs from platforms, most companies use platforms for production of oil and gas rather than for drilling. Because this book concentrates on drilling and not platforms, more information about platforms is available in another PETEX publication: *A Primer of Offshore Operations*.

If a platform is designed for drilling, the rig on the platform operates just like a land rig. Several wells can be drilled from the same platform, and the rig is moved or *skidded* over to the next slot in the platform to begin a new well.

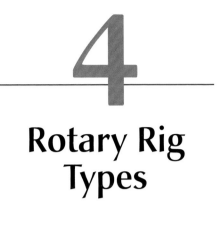

Rotary Rig Types

Figure 26. An offshore jackup rig

Figure 25. A land rig

Figure 27. An inland barge rig

LAND RIGS

Land rigs might look alike but individual specifications differ depending on the unique requirements of the site. Land rigs vary in size depending on how deep the rig can drill. Well depths can range from a few hundred or thousand feet (metres) to tens of thousands of feet (metres). The depth of the oil and gas formation targeted controls well depth. Land rigs are classified by size: light duty, medium duty, heavy duty, and very heavy duty. Table 1 arranges them according to this scheme and shows the depths to which they can drill.

Table 1
Land Rigs Classified by Drilling Depth

Rig size	Maximum Drilling Depth
Light duty	3,000–5,000 ft (1,000–1,500 metres)
Medium duty	5,000–10,000 ft (1,500–3,000 metres)
Heavy duty	10,000 – 16,000 ft (3,000–5,000 metres)
Very heavy duty	16,000–25,000+ ft (5,000–7,500+ metres)

A rig can drill holes shallower than its maximum rated depth. For example, a medium-duty rig could drill a 2,500-foot (750-metre) hole, although a light-duty rig could also drill the same depth. On the other hand, a rig cannot drill too far beyond its rated maximum depth because it cannot handle the heavier weight of the drilling equipment required for deeper holes.

A major feature of land rigs is that they are portable. A rig can drill a hole at one site, be disassembled, and then reassembled at another site to drill another hole (fig. 28). Rigs are so mobile that one definition terms them portable hole factories.

Figure 28.   Rigs can be disassembled and moved piece-by-piece to a new location.

MOBILE OFFSHORE RIGS

Today, about one-third of the world's oil comes from beneath the sea. To reach offshore oil, a *mobile offshore drilling unit*, or *MODU* is used. MODUs are portable. They are used to drill a well at one offshore site and then moved to drill another well. MODUs are either floating or *bottom-supported* (fig. 29).

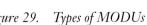

Figure 29.   Types of MODUs

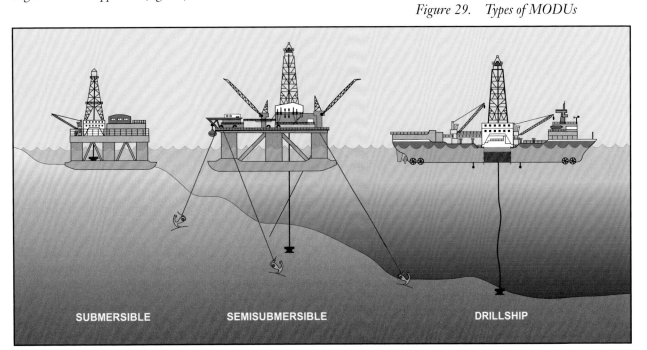

SUBMERSIBLE   SEMISUBMERSIBLE   DRILLSHIP

Floating systems do the work of drilling on top of, or slightly below, the water's surface. Floating systems include *semisubmersibles* and *drillships*, and they can drill in waters thousands of feet (metres) deep. Bottom-supported MODUs are in contact with, and are supported by, the ocean floor, and generally drill in water shallower than the water that floaters drill in. Bottom-supported units include *submersibles* and *jackups*. Submersibles are further divided into *posted barges*, *bottle types*, *inland barges*, and *Arctic types*. Table 2 lists types of MODUs.

Table 2
Types of MODUs

Mobile Offshore Drilling Units (MODUs)	
Bottom-Supported Units	**Floating Units**
Submersibles	Semisubmersibles
Posted Barges	Drillships
Bottle Types	
Arctic Types	
Inland Barges	
Jackups	

Bottom-Supported MODUs

Submersibles and jackups contact the seafloor when drilling. The lower part of a submersible's structure rests on the seafloor. In the case of jackups, only the legs contact the seafloor.

Submersible MODUs

A submersible floats on the water's surface while it is moved from one drilling site to another. When it reaches the site, crewmembers flood compartments that submerge the lower part of the rig to the seafloor. With the base of the rig in contact with the ocean bottom, wind, waves, and currents have little effect on it.

Posted-Barge Submersibles

The first MODU was a posted-barge submersible. The first well drilled by a posted barge was in 1949 off the Gulf Coast of Louisiana in 18 feet (5.5 metres) of water. The posted barge had a barge *hull* and steel columns that supported a deck and drilling equipment (fig. 30). This successful well proved that mobile rigs could drill offshore. Today, posted barges are virtually obsolete because better designs have replaced them.

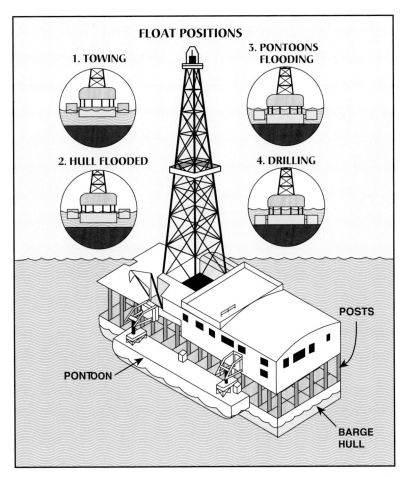

Figure 30.   The first MODU was a posted-barge submersible designed to drill in shallow water.

Bottle-Type Submersibles

In the mid 1950s, drilling moved into water depths of about 30 feet (9 metres), which is beyond the posted barge's capabilities. Naval architects designed bottle-type submersibles to drill in deeper water. A bottle-type rig has four tall steel cylinders (bottles) at each corner of the structure. The main deck lies across several steel supports and the bottles. The rig and other equipment are placed on the main deck. When flooded, the bottles cause the rig to submerge to the seafloor (fig. 31).

Until the early 1960s, the biggest bottle-type submersibles drilled in 150-foot (45-metre) water depths. Today, jackups have largely replaced bottle-type submersibles. Jackups are less expensive to build than bottle-type submersibles and can drill in deeper water. Rather than completely abandon bottle types, some rig owners modified their bottle types to become semisubmersibles.

Figure 31.   When the bottles are flooded, the weight makes the bottle-type rig sink to the seafloor.

SOURCE: NOAA, DEPARTMENT OF COMMERCE; PHOTOGRAPHER COMMANDER RICHARD BEHN, NOAA CORPS

Figure 32.   Ice floes on the North Bering Sea

Arctic Submersibles

A special type of submersible rig is the Arctic submersible. In the Arctic, where petroleum deposits lie under shallow oceans such as the Beaufort Sea, jackups and conventional barge rigs cannot be used. During the Arctic winters, massive chunks of ice moving with ocean currents form on the water's surface. Called *floes*, these moving ice blocks apply tremendous force on any object that they contact. The force is great enough to destroy the legs of a jackup or the hull of a conventional ship or barge (fig. 32). Therefore, Arctic submersibles have a reinforced hull called a *caisson*. One type of caisson has a reinforced concrete base on which the drilling rig is installed (fig. 33).

Figure 33.   A concrete island drilling system (CIDS) features a reinforced concrete caisson.

During the short Arctic summer when the sea is ice free, ships tow the submersible to the drilling site. On location, workers submerge the caisson to the sea bottom and start drilling. When ice floes form and begin to move, the Arctic submersible's strong caisson hull deflects the floes, allowing operations to continue.

Inland Barge Rigs

An inland barge rig has a barge hull—a flat-bottomed, flat-sided, rectangular steel box. The rig builder places a drilling rig and other equipment on the barge deck. Inland barge rigs normally drill in marshes, bays, swamps, or other shallow inland waters (fig. 34).

Barges are not *self-propelled*. They have no built-in power to move them from one site to another. Therefore, a boat must tow an inland barge to the drilling location. While being moved, the barge floats on the water's surface. Then, when positioned at the drilling site, the barge is flooded so it rests on the bottom of the water. Because inland barges drill in swampy shallow waters, drilling workers often call them *swamp barges*. In swampy locations, a channel or canal is dredged first so the rig can be positioned.

Figure 34.   Drilling equipment is placed on the deck of a barge to drill in the shallow waters of bays and estuaries.

Figure 35.   Four boats tow a jackup rig to its drilling location.

Jackups

A jackup rig is a widely used MODU that floats on a barge hull when towed to the drilling location (fig. 35). Many modern jackups have three legs with a triangular-shaped barge hull. Others have four or more legs with rectangular hulls. A jackup's legs can be cylindrical columns, similar to pillars (fig. 36). Or, they can be open-truss structures that resemble a mast or derrick (fig. 37).

Once the jackup's barge hull is in position at the drilling site, the crew jacks down the legs until they rest on the seafloor. Then they raise, or jack up, the hull above the height of the highest anticipated ocean waves in the location (fig. 38). The drilling equipment is on top of the hull. The largest jackups can drill in water depths up to about 400 feet (about 120 metres) and drill holes up to 30,000 feet (9,100 metres), or approximately 5½ miles, deep.

Figure 36.   A jackup rig with four column-type legs

Figure 37. A jackup with open-truss legs

Figure 38. The hulls of these jackups are raised to clear the highest anticipated waves.

Floating Units

Floating *offshore drilling rigs* include semisubmersibles and drillships. Because of their design, semisubmersibles are more stable than drillships. However, drillships can carry more drilling equipment and supplies. This makes drillships the preferred choice for drilling in remote waters.

Semisubmersibles

Most semisubmersible rigs have two or more *pontoons* on which the rig floats. A pontoon is a long, relatively narrow, hollow steel float with a rectangular or round cross section (fig. 39). When a semisubmersible is moved, the pontoons that contain mainly air allow the rig to float on the water's surface. Usually, towboats tie onto the rig and move it to the new drill site.

Another method of transporting the semisubmersible is to use a *heavy lift vessel*. A heavy lift vessel can partly submerge, allowing the semisubmersible to be floated over the deck. The heavy lift vessel then rises above the surface, lifting the semisubmersible and sails with the rig to the next location. This is commonly done when moving the rig a long distance. Some semisubmersibles are self-propelled;

Figure 39.   A semisubmersible rig floats on pontoons.

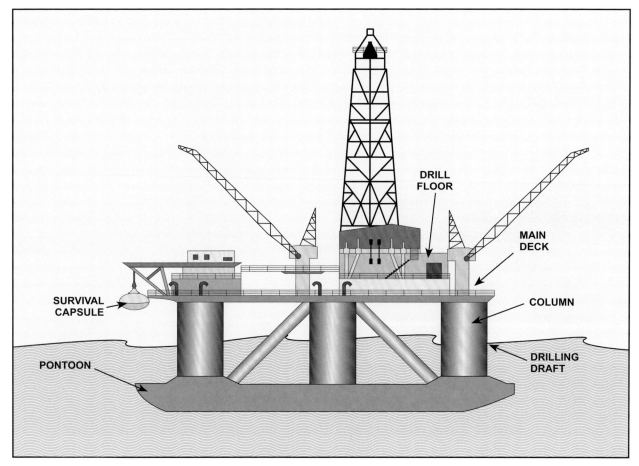

Figure 40.   The heavy lift vessel, Blue Marlin, transporting BP's semisubmersible, Thunder Horse

they have built-in power units that drive the rig from one site to another. For long moves, even self-propelled semisubmersibles often use a heavy lift vessel to save time (fig. 40).

While drilling, semisubmersibles are not *submerged* to the point where it makes contact with the sea bottom. Instead, rig personnel carefully flood the pontoons to make them submerge only a few feet (metres) below the water's surface (fig. 41). The rig is only semisubmerged, or partly submerged, resulting in the name semisubmersible.

Figure 41.   The pontoons of this semisubmersible float a few feet (metres) below the water's surface.

Waves do not affect the rig as much as they would if it were floating on the surface. Therefore, a semisubmersible rig offers a more stable drilling platform than a drillship that drills while floating on the water's surface.

Semisubmersibles have large cylindrical or square columns that extend upward from the pontoons (fig. 42). The main deck is huge and rests on top of the columns. Semisubmersibles often use anchors to keep them on the drilling station. Workers release several large anchors from the deck of the rig. An anchor-handling boat crew sets the anchors on the seafloor. Some semisubmersibles have *thrusters* that keep the rig dynamically positioned over the well without the use of anchors.

Besides being more stable in rough water, semisubmersibles can drill in water thousands of feet (metres) deep. While many semisubmersibles work in water depths ranging from 1,000 to 3,500 feet (300 to 1,000 metres), some can drill in water depths of 12,000 feet (3,660 metres). Semisubmersibles can drill holes up to 35,000 feet (10,670 metres) deep. They are among the largest floating structures ever constructed. The biggest ones are well over 100 feet (30 metres) tall with main decks sometimes over 3,000 square yards (2,500 square metres).

Figure 42.   The main deck of a semisubmersible is huge. Shown here is the deck of the BP Thunder Horse.

Drillships

A drillship is another type of floating system, or *floater*. Drillships are highly mobile; they are self-propelled and have a streamlined hull, much like most other oceangoing ships. A drillship is a preferred choice for more remote drilling locations. It can move at sufficient speed under its own power. Also, the ship-shaped hull can carry the large amount of equipment and material required for remote drilling, enabling less frequent resupplying from a shore base (fig. 43).

While drillships usually operate in water depths ranging from 1,000 to 3,000 feet (300 to 1,000 metres), many can drill in water depths of 10,000 feet (3,000 metres), or nearly 2 miles (3.2 kilometres). They can drill holes over 30,000 feet (9,100 metres) deep. These big drillships are more than 800 feet (250 metres) long and they measure approximately 100 feet (30 metres) wide. Their hulls are more than 60 feet (18 metres) high, about the size of a six-story building.

COURTESY OF TRANSOCEAN

Figure 43.   Deepwater Pathfinder 10,000-foot ultradeepwater drillship

Anchors keep some drillships on station during drilling, but those drilling in deep water require a computer-controlled *dynamic positioning system*. Dynamically positioned drillships use computerized thrusters and advanced technology sensors. The thruster power units have propellers in the front (*fore*) and back (*aft*) on the drillship's hull below the waterline. The *dynamic positioning operator* programs the computer to keep the rig positioned. The computer uses information transmitted by sensors and automatically controls the thrusters. The thrusters counteract wind, wave, and ocean current forces that would otherwise move the rig away from the desired position.

The floating rigs are different from the bottom-supported rigs in the way the floater is connected to the well at the seafloor. On a floater, the *blowout preventers* are placed on the seafloor and anchored to the well casing. The blowout preventers control the well and are connected to the floater by a long segmented pipe called a *marine riser* (fig. 44).

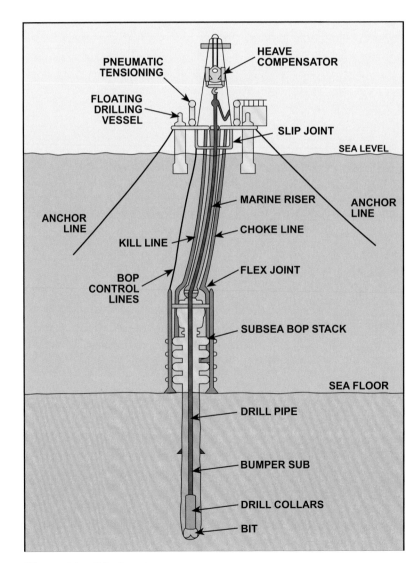

Figure 44.   Marine riser

The marine riser acts as a continuation of the well, allowing circulation of the drilling fluid. When the well is finished, the blowout preventers are removed and a *production wellhead* is installed in its place.

Divers can install the equipment on the seafloor as long as the water is not too deep. If the required depth is too deep to use divers, a *remotely operated vehicle (ROV)* is used to view the installation and assist in installing the equipment on the seafloor. Bottom-supported rigs can have the blowout preventers installed on top of the well casing extended above sea level to the deck of the rig.

Floating rigs are also affected by waves and ocean swells, whereas the bottom-supported rigs are not. Floaters must have special *tensioners* and ball joints for the marine riser and *heave compensators*. Heave compensators located in the derrick allow the drill string to remain stationary and unaffected by wind and waves (figs. 45a and 45b).

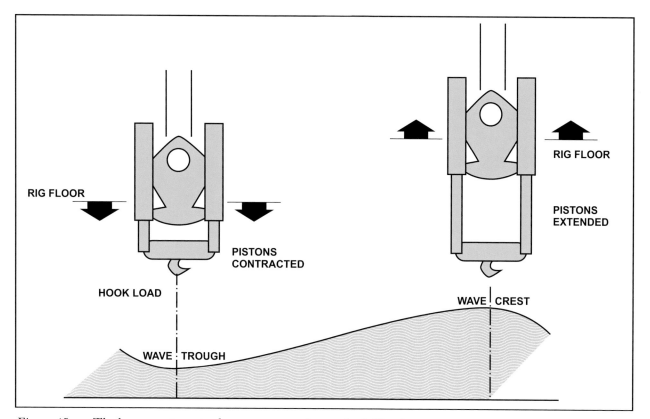

Figure 45a.   The heave compensator keeps proper tension on the drill string.

Figure 45b.   Heave compensator

COURTESY OF IODP-USIO

W hether on land or offshore, and regardless of size, all rigs require personnel to operate them. There are people employed by companies involved in drilling work all over the world. They drill wells on land and ice, in swamps, and on water as small as lakes or as large as the Pacific Ocean. Drilling is demanding work, continuing 24 hours a day, 7 days a week, in all kinds of weather (fig. 46).

Drilling is also increasingly complex. The technical complexity is so great that no single company is diverse enough to perform all the required work. Consequently, many companies and individuals are involved in drilling a well, including operating companies, drilling contractors, and service and supply companies.

5

People and Companies

Figure 46.   Workers on a drilling rig

OPERATING COMPANIES

An *operating company*, or an *operator*, is usually an oil company. This company's primary business is working with oil and gas, known as petroleum. An operating company may be termed either an *independent* or a *major*. An independent company might be one or two individuals, or it might have hundreds of employees. Major companies might have thousands of employees. In general, an independent only produces and sells crude oil and natural gas. A major produces crude oil and natural gas, transports them from the field to a refinery or a plant, refines or processes the oil and gas, and sells the products to consumers.

Whether an independent or a major, the operator must acquire the right to drill for and produce petroleum at a particular site. An operating company usually does not own the land or offshore tract, or the *mineral rights* (oil and gas are considered minerals) lying under the surface. The company has to buy or *lease* the rights to drill for and produce oil and gas from the land or tract owner and the mineral holder. Individuals, partnerships, corporations, or a federal, state, or local government can own land or offshore tracts and mineral rights (fig. 47).

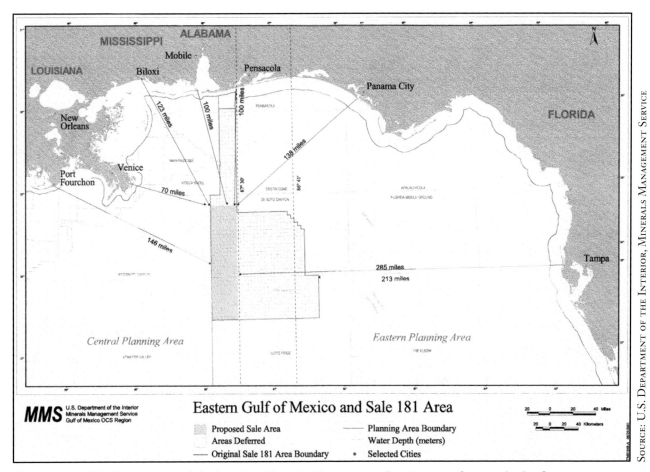

SOURCE: U.S. DEPARTMENT OF THE INTERIOR, MINERALS MANAGEMENT SERVICE

Figure 47. U.S. Department of the Interior, Minerals Management Service map of proposed sale of government mineral leases in 2001

The operator not only pays the landowner a fee for leasing, it also pays the mineral holder a *royalty*, which is a share of the money made from the sale of oil or gas. Companies that operate in countries other than the United States must make similar lease or concession agreements with the governments of the countries. Frequently, a country's government becomes a partner in the operation.

DRILLING CONTRACTORS

Drilling is an undertaking that requires experienced personnel and special equipment. Most operating companies find it cost effective to hire expertise and equipment from *drilling companies* rather than employ personnel and purchase equipment themselves. Almost everywhere in the world, it is the drilling contractors who do the actual drilling work.

A *drilling contractor* is an individual or a company that owns one or a large number of drilling rigs. The drilling contractor *contracts* with operators to provide a rig and the personnel needed to operate it. Some are land contractors who operate only land rigs; others are offshore contractors who operate only offshore rigs. A few contractors operate rigs that drill both on land and offshore. The contractor might have different sizes of rigs that can drill to various depths. A drilling contracting company could be small or large, owning rigs that drill mainly in local areas or rigs working all over the world.

Regardless of size, a drilling contractor's job is to drill holes. The drilling contractor will drill holes to the depth and specifications set by the operating company that owns the well. An operating company usually asks several contractors to bid their costs to do a job, and the operator often awards the contract to the lowest bidder. Sometimes a good work record overrides a low bid.

DRILLING CONTRACTS

The operator usually sends a proposal to several drilling contractors. The proposal describes the drilling project and requests bids for the work. Most companies and contractors employ their own lawyers to review and approve all contracts. If interested in the project, the contractor writes the proposal, signs it, and sends it back to the operator. Once the operating company accepts the bid and signs the proposal, there exists a legal contract between the operator and the drilling company. This signed agreement clearly states the services and supplies the contractor and the operator are to provide for a particular project. A *drilling contract* will also specify a term or time period.

The *International Association of Drilling Contractors (IADC)* supplies standard contract forms that can be modified for a specific project by the company and contractor's legal office (fig. 48). IADC is an organization headquartered in Houston, Texas, with a membership made up of drilling contractors, oil companies, and service and supply companies involved in drilling.

Drilling contractors are paid for the work their rig and crews provide in several ways. Operators can pay contractors based on the daily costs of operating the rig, the number of feet or metres drilled, or on a *turnkey* basis. If the contractor is paid according to the daily costs of operating the rig, it is called a *daywork contract*. A daywork contract specifies that the operator will be responsible for all activities and costs while drilling the well. If the contract calls for the contractor to be paid by the number of feet or metres drilled, it is a *footage rate* or *metreage contract*. A footage contract requires that the contractor be responsible for the costs and activities associated with the rig, and the operator is responsible for other items such as the drilling fluid, site preparation, and well casing. When the operator and contractor sign a *turnkey contract*, the drilling contractor is responsible for the entire drilling operation and is paid a specified fee when the job is completed. Daywork contracts are the most common.

SERVICE AND SUPPLY COMPANIES

To profitably drill a well, the operator and the drilling contractor must have equipment, supplies, and services that neither company has available. Service and supply companies are used to acquire the necessary tools and services to expedite well drilling.

Supply companies sell expendable and nonexpendable equipment and material to the operator and the drilling contractor. Expendable items include drill bits, fuel, lubricants, and drilling mud—items that are used up or worn out as the well is drilled. Nonexpendable items include drill pipe and other tubular well products, wellheads, and equipment that might eventually wear out and have to be replaced but normally last a long time. Supply companies also market safety equipment, rig components, tools, computers, paint, grease, rags, and solvents. Any part or commodity that a rig needs to drill a well comes from a supply company.

Service companies offer special support to the drilling operation. For example, a *mud logging* company provides a service to monitor and record, or *log*, the content of the drilling mud as it returns from the well. The returning mud carries cuttings and formation fluids, such as gas or oil, to the surface. The operator can gain knowledge about the formations being drilled from the log.

NOTE: This form contract is a suggested guide only and use of this form or any variation thereof shall be at the sole discretion and risk of the user parties. Users of the form contract or any portion or variation thereof are encouraged to seek the advice of counsel to ensure that their contract reflects the complete agreement of the parties and applicable law. The International Association of Drilling Contractors disclaims any liability whatsoever for loss or damages which may result from use of the form contract or portions or variations thereof.

Revised April, 2003

INTERNATIONAL ASSOCIATION OF DRILLING CONTRACTORS
DRILLING BID PROPOSAL
AND
FOOTAGE DRILLING CONTRACT - U.S.

TO: _____

_____

Please submit bid on this drilling contract form for performing the work outlined below, upon the terms and for the consideration set forth, with the understanding that if the bid is accepted by _____ this instrument will constitute a Contract between us. Your bid should be mailed or delivered not later than _____ P.M. on _____, 20_____ to the following address: _____

THIS CONTRACT CONTAINS PROVISIONS RELATING TO INDEMNITY, RELEASE OF LIABILITY, AND ALLOCATION OF RISK - SEE PARAGRAPHS 4.7, 6.2, 15, AND 19

This **Contract** is made and entered into on the date hereinafter set forth by and between the parties herein designated as "Operator" and "Contractor".

OPERATOR: _____

Address: _____

CONTRACTOR: _____

Address: _____

IN CONSIDERATION of the mutual promises, conditions and agreements herein contained and the specifications and special provisions set forth in Exhibit "A" and Exhibit "B" attached hereto and made a part hereof (the "Contract"), Operator engages Contractor as an independent contractor to drill the hereinafter designated well in search of oil or gas on a Footage Basis.

For purposes hereof, the term "Footage" or "Footage Basis" means Contractor shall furnish the equipment, labor, and perform services as herein provided to drill a well, as specified by Operator, to the Contract Footage Depth. Subject to terms and conditions hereof, payment to Contractor at a stipulated price per foot of hole drilled is earned upon attaining such Contract Footage Depth or other specified objective. While drilling on a Footage Basis Contractor shall direct, supervise and control drilling operations and assumes certain liabilities to the extent specifically provided for herein. Notwithstanding that this is a Footage Basis contract, Contractor and Operator recognize that certain portions of the operations as hereinafter designated, both above and below Contract Footage Depth, will be performed on a Daywork Basis. For purposes hereof, the term "Daywork" or "Daywork Basis" means Contractor shall furnish equipment, labor, and perform services as herein provided, for a specified sum per day under the direction, supervision and control of Operator (inclusive of any employee, agent, consultant, or subcontractor engaged by Operator to direct drilling operations). *When operating on a Daywork Basis, Contractor shall be fully paid at the applicable rates of payment and assumes only the obligations and liabilities stated herein as being applicable during Daywork operations. Except for such obligations and liabilities specifically assumed by Contractor, Operator shall be solely responsible and assumes liability for all consequences of operations by both parties while on a Daywork Basis, including results and all other risks or liabilities incurred in or incident to such operations.*

1. **LOCATION OF WELL:**
    Well Name and Number: _____
    Parish/ County: _____ State: _____ Field Name: _____
    Well location and land description: _____
    _____

The above is for well and Contract identification only and Contractor assumes no liability whatsoever for a proper survey or location stake on Operator's lease.

2. **COMMENCEMENT DATE:**
    Contractor agrees to use reasonable efforts to commence operations for the drilling of the well by the _____ day of _____, 20_____, or _____

3. **DEPTH:**
    Subject to the provisions hereof, the well shall be drilled to the depth as specified below:

    3.1 Contract Footage Depth: The well shall be drilled to _____ feet or formation, or to the depth at which the _____ inch casing or liner is set, whichever depth is first reached, on a Footage Basis and Contractor is to be paid for such drilling at the footage rate specified below, which depth is herein referred to as the Contract Footage Depth.

    3.2 Daywork Basis Drilling: All drilling below the above specified Contract Footage Depth shall be on a Daywork Basis as defined herein and Contractor shall be paid for such drilling at the applicable Daywork rate specified in Paragraph 4.

    3.3 Complete Daywork Basis Drilling: If all operations hereunder are performed at applicable Daywork rates, provisions of this Contract applicable to drilling on a Footage Basis shall not apply.

    3.4 Maximum Depth: Contractor shall not be required to drill said well under the terms of this Contract below a maximum depth of _____ feet.

4. **FOOTAGE RATE, DAYWORK RATES, BASIS OF DETERMINING AMOUNTS PAYABLE TO CONTRACTOR:**
    Contractor shall be paid at the following rates for the work performed hereunder.

    4.1 Footage Rate: For work performed on a Footage Basis the rate will be $_____ per linear foot of hole drilled determined by steel line measurement from the surface of the ground if Contractor provides cellar, or from the bottom of the cellar if Operator provides cellar, less footage made in regular size hole while working on Daywork Basis.

    4.2 Operating Rate: For work performed on a Daywork Basis the operating rate per twenty-four hour day with _____ man crew shall be:

Depth Intervals		Without Drill Pipe	With Drill Pipe
From	To		
_____	_____	$_____ per day	$_____ per day
_____	_____	$_____ per day	$_____ per day
_____	_____	$_____ per day	$_____ per day

Using Operator's drill pipe $_____ per day.

(U.S. Footage Contract - Page 1)
Copyright © 2003 International Association of Drilling Contractors

COURTESY OF IADC

Figure 48.   IADC standard drilling bid form. When signed by all parties, it becomes a legal contract.

Often, when a well reaches a formation that might contain oil or gas, the operating company hires the services of a *well logging* company. A logging crew runs sophisticated instruments into the hole to sense and record formation properties. Computers in the field generate graphs, called *well logs*, for the operator to examine (fig. 49). Well logs help the operating company determine whether the well will produce oil or gas.

Figure 49.   A computer display showing a well log

Another service company provides *casing crews*. The casing crew runs pipe called casing to line the well after the rig drills a portion of the hole (fig. 50). The casing protects formations from environmental contamination during drilling and stabilizes the well. After the casing crew runs the casing, another service company—a *cementing company*—cements the casing in the well. *Cementing* bonds the casing to the hole.

There are also companies that provide catering and housekeeping services on offshore rigs and on land rigs in remote areas. Because the workers live offshore or in isolated regions for long periods, the company contracts with an oilfield caterer and janitorial companies to furnish these services (fig. 51).

CASING COUPLING

Figure 50.   A member of a casing crew stabs one joint of casing into another. The red fitting is a casing coupling, called a collar, used to connect the joints.

Figure 51.   Personnel on this offshore rig enjoy quality food in the galley. A catering company usually provides the food and cooking.

PEOPLE

Drilling a well requires many people to run the rig, and keep it running, until the well reaches its goal. A crew of people with diverse skills is needed to accomplish any drilling project.

Drilling Crews

The contractor must have well-trained, skilled personnel to operate and maintain a drilling rig. When on site and drilling, a rig operates virtually nonstop, night and day, 365 days a year. The *drilling crew* is the group of personnel whose job it is to make the rig actually drill.

The person in charge of the drilling crew might be called a rig manager, rig superintendent, or toolpusher, depending on the drilling contractor's preference. In addition to the rig manager, or superintendent, each rig has drillers, *derrickmen*, and *floorhands* (also called *rotary helpers*, or *roughnecks)*. Large land rigs and offshore rigs often have assistant *rig supervisors*, *assistant drillers*, and other personnel to perform special functions specific to the rig.

Rig Superintendent and Assistant Rig Superintendent

The rig superintendent (rig manager or toolpusher) oversees drilling crews working on the rig floor, supervises drilling operations, and coordinates operating company and drilling contractor interactions. On land rigs, the rig superintendent is usually headquartered in a mobile home or a portable building at the rig site and is on-call all the time. Offshore, the rig superintendent has an office and sleeping quarters on the rig and is also always on-call.

Because offshore drilling and large land drilling operations can be extremely demanding, the contractor might hire an *assistant rig superintendent*. The assistant rig superintendent frequently relieves the superintendent during nighttime hours and is sometimes nicknamed the *night toolpusher*.

Driller and Assistant Driller

The rig superintendent supervises the driller. The driller, in turn, supervises the assistant driller, derrickman, and floorhands. From a *driller's console* or an operating cabin on the rig floor, the driller manipulates the controls that keep the drilling operation going (fig. 52).

The driller is directly responsible for drilling the hole. Most offshore rigs and large land rigs, especially those working outside the United States, also have an assistant driller. The assistant driller helps the driller on the rig floor and assists in supervising the derrickman and floorhands.

Figure 52.   A driller on an offshore rig works in an environmentally controlled cabin.

Derrickman

Some rigs now have automated pipe-handling equipment that takes over some of the derrickman's duties. Most rigs still need a derrickman when crewmembers run drill pipe into the hole (*tripping in*), or when they pull pipe out of the hole (*tripping out*). The derrickman handles the upper end of the pipe from the *monkeyboard* (fig. 53).

The monkeyboard is a small platform high on the derrick on which the derrickman stands to handle the upper end of the pipe. The derrickman mounts the monkeyboard at a height ranging from about 50 to 110 feet (15 to 34 metres), depending on the length of the pipe joints that crewmembers pull from the hole.

Figure 53.   The view from above the derrickman's position on the monkeyboard. Drill pipe is racked to the left. Drilling line runs from the top to the traveling block below. Note that this rig has eight lines strung.

Special safety equipment is used by the derrickman to prevent falls and injury. In addition, the derrickman has an escape device, called a *Geronimo*, or *Tinkerbell line*[1], to emergency exit the monkeyboard. To get out of the derrick quickly, the derrickman grasps a handle on the Geronimo and rides it down on a special cable, or line, to the ground. The derrickman controls the rate of descent by moving the handle to increase or decrease the braking action on the line.

When the pipe is in the hole, the derrickman, using a built-in ladder in the derrick, climbs down from the monkeyboard and works at ground level. The derrickman is responsible for any work done in the derrick including maintenance of the *crown block*. When not in the derrick, the derrickman monitors the condition of the drilling mud (fig. 54). The derrickman also ensures the drilling mud meets specifications for drilling a particular part of the hole by mixing in the mud additives specified by the *mud engineer*.

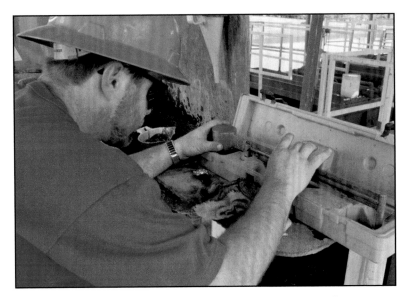

Figure 54.   A derrickman checking the weight or density of the drilling mud

---

[1] Geronimo was a Chiricahua Apache who eluded the Army for many years in the American southwest in the late 1800s. World War II paratroopers sometimes yelled his name as they jumped out of airplanes. Tinkerbell is a fictitious flying character from the children's novel *Peter Pan*.

Figure 55.   Floorhands latch big wrenches called tongs onto the drill pipe.

Floorhands (Rotary Helpers or Roughnecks)

A contractor usually hires two or three floorhands, or rotary helpers, for each work shift depending on the size of the rig, its equipment, and other factors. Floorhands get their name from the fact that much of their work occurs on the rig floor, near the rotary table—the device that turns the drill pipe and bit. Originally, all the members of a drilling crew, including the floorhands, were called roughnecks, probably because those who worked on early rigs prided themselves on being rough and tough.

Floorhands handle the lower end of the drill pipe when tripping it in or out of the hole. They also use large wrenches called *tongs* to screw or unscrew (*make up* or *break out*) pipe (fig. 55). At times, power tongs rather than manual tongs are used (fig. 56).

Figure 56.   Floorhands using power tongs to tighten drill pipe

Beside handling pipe, floorhands also maintain the drilling equipment, help repair it, and keep it clean and painted. On small rigs drilling shallow wells, for example, two floorhands on a shift can safely and efficiently perform the required duties. On offshore rigs and large land rigs drilling deep holes, the job usually requires three floorhands.

A new worker will start out as a floorhand and can progress through the ranks to become a derrickman, a driller, a toolpusher, and an area rig superintendent. Each progression requires workers to master new skills, and they are rewarded with increasing responsibility and pay. Drilling crewmembers are often paid by the hour, although some might be paid by the day or the rotation. Toolpushers and area drilling superintendents are frequently paid a set monthly salary. Benefits and bonuses for performance and job safety are common.

Drilling Crew Work Shifts

The number of days and the number of hours per day a drilling crew works varies greatly depending on a rig's geographic location and economic and other factors. Drilling crews call their shifts *tours*[2]. In a few areas, particularly in West Texas and Eastern New Mexico, contractors use 8-hour tours. In other areas, such as offshore, along the Gulf Coast, in countries outside the United States, and in remote land locations, they use 12-hour tours.

When crews work 8-hour tours, the contractor usually hires four drilling crews and two toolpushers, or rig superintendents, for each rig. While three drilling crews split three 8-hour tours per day, a fourth crew is off duty. Later, the fourth crew relieves one of the working crews. One rig superintendent, or toolpusher, is on the site at all times. He or she might work 7 days and then be relieved by the other superintendent for 7 days.

If the crews work 12-hour tours on land, then the contractor might hire three drilling crews and two superintendents for each rig: two full drilling crews—split two tours per 24-hour day; the third crew is off duty and rotates with one of the working crews to allow days off. This system also allows the crews to rotate from working days to working nights so no single crew is always on the night shift.

Offshore crews usually work 12-hour tours, but the contractor hires four drilling crews. Two crews might work 14 days and then take 14 days off when the second crews come on board to relieve them. Some contractors based in the United States have rigs working abroad, such as in the North Sea or in Southeast Asia. In such cases, the contractor often employs a 28-and-28 schedule. Two crews are home for 28 days while the other two work 12-hour tours for 28 days.

Crew Safety

On-the-job safety is an important part of the daily routine. Each crewmember must wear personal protection equipment (PPE). PPE includes shatter-proof safety glasses, a hard hat, steel-toed boots, work gloves, coveralls, and ear plugs. Safety training is provided for all crewmembers regularly. A safety meeting (sometimes called a *value moment*) is held before the beginning of each tour and before the start of any specialty job. Offshore, crew safety training includes mandatory helicopter and crew boat safety and abandon-ship drills. Crewmembers are taught to look out for each other and to point out any unsafe practices or equipment. Safety bonuses of cash or prizes are awarded for crews that maintain good safety records.

---

[2] Tours is pronounced as like the word "towers."

Other Rig Workers

Besides the drilling crew, many other workers are at the rig site during drilling or they are at the rig only while their expertise or equipment is needed. All workers are required to meet the same safety standards as the drilling crew.

The Company Representative

The operating company customarily has an employee on the drill site to supervise company interests. On a land rig, the *company representative*, or *company man*, usually lives on the rig site in a mobile home or portable building. Offshore, the company representative has an office and designated quarters. Either on land or offshore, the company representative is in charge of all the operator's activities on the location. This person helps plan the strategy for drilling the well, orders the needed supplies and services, and makes on-site decisions that affect the well's progress. The company representative and the rig superintendent usually work together closely.

Area Drilling Superintendent

Large land drilling contractors operating rigs worldwide or in one area often employ an *area drilling superintendent*. This person's job is to manage and coordinate the activities of the drilling company's rigs working in a particular region. An area superintendent's duties include distributing important information to each of the rigs, ensuring all rigs are operating well and safely, and assisting each rig's superintendent as required. Because area drilling superintendents frequently travel from rig to rig, they generally have an office in a town or city in the local area.

Offshore Personnel

The sea and remoteness of the site complicate operations offshore. The offshore operating contractor requires more personnel than on land. Regulations in many areas require offshore rigs to have an *offshore installation manager (OIM)*. The OIM is in charge of the entire rig and has final say in any decision affecting the rig's operation. On some offshore rigs, the rig superintendent also serves as the OIM, while other offshore rigs might have both an OIM and a rig superintendent.

Figure 57. Roustabouts move casing from a supply boat to the rig.

Offshore contractors also hire several *roustabouts*. Roustabouts are general rig workers whose duties include unloading supplies from boats to the rig (fig. 57). They are also responsible for the maintenance of the offshore facilities.

A *crane operator* runs the rig's cranes and supervises the roustabouts (fig. 58). Cranes transfer supplies to and from boats. Radio operators install, maintain, and repair complex radio gear that keeps the rig in constant contact with shore facilities. Medics provide first aid and are often certified emergency medical technicians (EMT) who can stabilize injured personnel and prepare them for evacuation to shore.

On floating rigs, such as drillships and semisubmersibles, marine personnel are required because the floating rigs operate much like ships. Floating rigs not only drill, but they also move on the ocean's surface. Consequently, floaters require *marine crews* with individuals responsible for the seagoing aspects of the rig.

Floating rigs also require subsea equipment. Crewmembers place the equipment on the seafloor and operate it from the rig on the water's surface. This equipment includes *subsea blowout preventers*. When closed, these large valves keep high-pressure fluids from escaping to the surface. To monitor these systems, floating rigs employ *subsea equipment supervisors*, also called *subsea engineers*, whose primary job is to keep the equipment in working order and supervise its installation on the seafloor. Often, floaters also have an assistant subsea equipment supervisor.

Figure 58.   A crane operator manipulates controls from a position inside the crane cab.

Floating offshore rigs also employ *barge control operators*, also called *barge masters* or *barge engineers*. Semisubmersible rigs with pontoon-shaped hulls submerged just below the water's surface require barge engineers to keep the rig stable and balanced while at work or being moved (fig. 59).

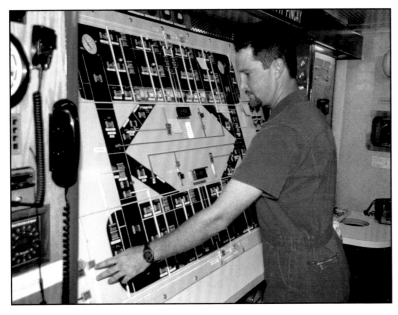

Figure 59.   A barge engineer monitors a semisubmersible's stability from a work station on board the rig.

This job is important because the rig must be kept stable. If the rig is allowed to tilt or list to one side, it might overturn and sink. Figure 60 shows a rig whose semisubmersibles pontoons were not kept correctly filled with water and air. The BP Thunder Horse rig was evacuated during a severe storm, so the barge operator was not on board to keep the rig *trim* or balanced resulting in a near-loss of the platform. The list was corrected after the crew returned.

COURTESY OF BP AMERICA INC.

Figure 60.   BP's Thunder Horse listing in the Gulf of Mexico after a storm

Office Personnel

Vital to any drilling project are the people who support the company offices. Operating companies, drilling contractors, and service and supply companies hire geologists, accountants, bookkeepers, sales personnel, and trainers. They also hire personnel specialists, planners, drilling engineers, environmental specialists, warehouse personnel, and safety specialists. In addition, they employ truck drivers, storage yard personnel, lawyers, drafting technicians, and a clerical staff to back up those in the field. Without a competent office staff, no company or contractor can maintain drilling operations.

Oil and gas are naturally occurring *hydrocarbons*. Two elements, hydrogen and carbon, make up a hydrocarbon. Because hydrogen and carbon have a strong attraction for each other, they form many compounds. The oil industry processes and refines crude hydrocarbons recovered from the earth to create hydrocarbon products including: natural gas, *liquefied petroleum gas* (LPG, or *hydrogas*), gasoline, kerosene, diesel fuel, and a vast array of synthetic materials such as nylon and plastics.

Crude oil and natural gas occur in tiny openings of buried layers of rock. Occasionally, the crude hydrocarbons ooze to the surface in the form of a seep, or spring. More often, rock layers trap the hydrocarbons thousands of feet (metres) below the surface. To bring the trapped hydrocarbons to the surface, operating companies and drilling contractors drill wells.

6
Oil and Gas: Characteristics and Occurrence

NATURAL GAS

The simplest hydrocarbon is *methane* (CH_4). It has one atom of carbon (C) and four atoms of hydrogen (H). Methane is a gas under standard conditions of pressure and temperature. Standard pressure is the pressure the atmosphere exerts at sea level, about 14.7 psia (101 kPa). Standard temperature is 60 degrees Fahrenheit, or 15.6 degrees Celsius.

Methane is the main component of natural gas. Natural gas occurs in buried rock layers usually mixed with other hydrocarbon gases and liquids. It sometimes also contains nonhydrocarbon gases and liquids such as helium, carbon dioxide, nitrogen, water, and *hydrogen sulfide*. Hydrogen sulfide is a poison that has a detectible sour or rotten-egg odor, even in low concentrations. Natural gas that contains hydrogen sulfide is called *sour gas*. After natural gas is produced or recovered, a gas processing facility removes impurities so the gas can be used by consumers.

Liquefied Natural Gas (LNG)

Liquefied natural gas (LNG) should not be confused with liquefied petroleum gas (LPG). LNG is natural gas that is a liquid at near standard pressure and very cold temperature (fig. 61). LNG is created using a cryogenic process in which natural gas is cooled to –260°F (–162°C) at approximately standard pressure to liquefy it. Once the gas becomes a very cold liquid, it occupies about $\frac{1}{600}$th of the volume it occupied as a gas at standard conditions. Large volumes of LNG can be shipped safely in special LNG tankers to overseas markets that pipelines cannot reach. When LNG is warmed it becomes a gas.

COURTESY OF STATOIL HYDRO. PHOTOGRAPH BY ROAR LINDEFJELD

Figure 61.   Arctic Discoverer LNG transport ship

Liquefied Petroleum Gas (LPG)

Liquefied petroleum gas (LPG) is mainly propane (C_3H_8) and butane (C_4H_{10}). It might also contain ethane (C_2H_6). Ethane, propane, and butane, which are heavier than methane, often occur with natural gas. Gas processing equipment removes these products from methane. LPG is a heavy component of natural gas that is a liquid under higher-than-standard pressure and normal or standard temperature.

By raising the pressure on propane and butane slightly above atmospheric pressure at standard temperature, they liquefy. When the pressure is released, propane and butane turn into gas. As a result, LPG can be used as a portable fuel. It can be transported in a pressurized container as a liquid. When the container is connected to a stove burner, for example, the LPG changes into gas when the burner is turned on and the pressure is released. Most people are familiar with LPG either for use in homes where it is stored under pressure in metal bottles called propane tanks or in portable bottles for space heaters and stoves. LPG is the heavy components of natural gas that are a liquid under higher-than-standard pressure and normal or standard temperature.

Natural Gas Liquid (NGL)

Natural gas liquid (NGL) is in a gaseous state at reservoir temperatures and pressures. When the pressure and/or temperature are lowered, hydrocarbon liquids condense out of the gas. They are sometimes called natural gas *condensate*. Condensate is light with a low density and is often colorless. Unlike LNG and LPG, condensate stays a liquid at standard conditions. Condensate sometimes is called white gas and, with the addition of antiknock compounds, was used as fuel in early gasoline engines. Modern gasoline engines are not made to run on this relatively unrefined form of gasoline. Natural gas liquids can be refined to form gasoline or used as *petrochemical feedstock* in industrial processes.

CRUDE OIL

Crude oil is a hydrocarbon mixture that occurs as a liquid. Some crude oils are extremely thick and dense and do not flow easily. Crude oil varies considerably in density, viscosity, and color. It might also contain nonhydrocarbon impurities such as hydrogen sulfide, *sediment*, and small amounts of water. Generally, oil companies classify crude oil as light, intermediate, or heavy. It is called *sour crude* if it contains hydrogen *sulfide*. Crude oil that does not contain hydrogen sulfide is classified as *sweet crude*. The sulfur component of sour crude is difficult and expensive to remove through refining. Sweet crude is preferred because it is less costly to refine and generally sells at a higher price than sour crude oil.

REFINED HYDROCARBONS

Oil refineries put crude oil through several chemical and physical processes to turn it into useful products such as gasoline, kerosene, diesel fuel, and lubricating oil. The refined products are mixtures, or blends, of several hydrocarbons that are liquid under normal conditions. Gasoline is made up of liquid hydrocarbons that are lighter in weight (less dense) than the liquid hydrocarbons that make up kerosene, diesel fuel, and lubricating oil.

OIL AND GAS RESERVOIRS

Hydrocarbons and associated impurities occur in rock formations usually buried thousands of feet or metres below the surface. Rock formations that hold hydrocarbons are often called reservoirs by scientists and engineers.

Oil does not flow in underground rivers or pool in subterranean lakes. Crude oil and natural gas occur in buried rocks. After it is produced from a well, companies refine the crude oil and process the natural gas into useful products. Not every rock holds hydrocarbons. To serve as an oil and gas reservoir, rocks must have certain properties.

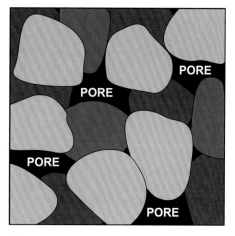

Figure 62. A pore is a small open space in a rock.

Characteristics of Reservoir Rocks

Nothing seems more solid than rock. However, *sandstone* or *limestone* actually have many tiny openings or *voids* in the rock. Geologists call these tiny openings *pores* (fig. 62).

A rock with pores or open spaces is called *porous*. Reservoir rocks must be porous because hydrocarbons can occur only in the pores. A porous rock has a measurable quality called *porosity*. Rock porosity is the measure of the rock's volume that is empty or a pore space. Porosity can be very low, almost zero. In theory, porosity can be as high as 55%, but that is only for extremely *well-sorted* (same diameter) rock grains. Because most rocks have grains of varying sizes, the practical limit of porosity is usually around 30% for sandstone (fig. 63).

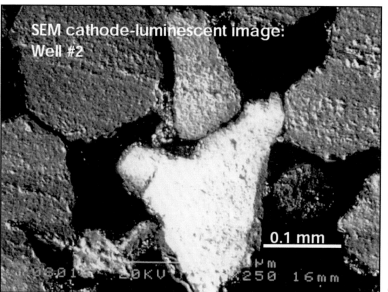

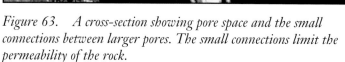

SOURCE: NATIONAL ENERGY TECHNOLOGY LABORATORY, U.S. DEPARTMENT OF ENERGY

Figure 63. A cross-section showing pore space and the small connections between larger pores. The small connections limit the permeability of the rock.

Porosity can also form when part of the rock dissolves. Limestone can be dissolved by water. The voids left in limestone are called *vugs*. Extremely large vugs can become caves. For example, the Edwards Water Aquifer in central Texas is made up of underground caves that were formed by the movement of water through the limestone. Fractures are another form of porosity. Porosity can be enhanced by chemical changes in the rock such as when limestone is converted to *dolomite* through the crystallization of calcium magnesium carbonate $CaMg(CO_3)_2$.

A reservoir rock must also be *permeable*—that is, its pores are connected so fluid can flow through the rock (fig. 64).

If hydrocarbons are in the rock pores, they must be able to move out of the pores. Unless hydrocarbons can move from pore to pore, they remain locked in place, unable to flow into a well. A suitable reservoir rock must be porous, permeable, and contain enough hydrocarbons to make it economically possible for the operating company to drill for and produce oil or gas. The higher the permeability, the easier it is for fluids to flow out of the rock.

Permeability is measured as a *darcy* (square metre). This measure is named after the French engineer Henry Darcy who in 1855–56 conducted pioneering experiments of liquid flowing through sand columns.

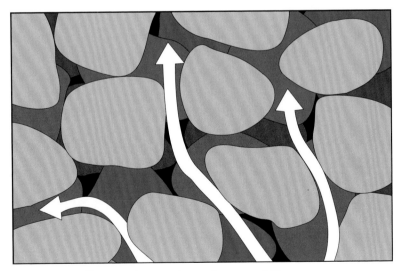

Figure 64.   Connected pores give rocks permeability.

Origin and Accumulation of Oil and Gas

Scientists theorize that hydrocarbons developed in rocks buried under ancient seas full of vast numbers of living organisms. Some creatures in the seas were fish and others were large swimming creatures. However, most organisms were so small they cannot be seen individually without magnification under a microscope. Although small, the microscopic creatures were extremely abundant.

Under the right conditions, large groups of these type organisms can be seen today in oceans and lakes. They appear as a green, green-blue, or brownish cloud in the water referred to as an algae bloom. Scientists believe these tiny organisms are the origin of oil and gas accumulations. Millions of years ago as these organisms died, the remains settled on the bottom of the water. Over thousands of years, enormous quantities of these dead organisms became organic sediment that accumulated in thick deposits on the seafloor.

The organic-rich sediment is known as *kerogen*. This organic material mixed with the mud and sand on the seafloor. Many layers of sediments built up and became hundreds or thousands of feet (metres) thick. The tremendous weight of the overlying sediments created great pressure and heat, changing the deep layers into rock. At the same time, heat, pressure, and other forces changed the kerogen into hydrocarbons: crude oil and natural gas.

Geologic forces slowly altered the shape of the earth creating cracks, or *faults*, in the earth's crust. Earth movement along the boundaries of *tectonic plates* folded layers of rock upward and downward. Molten rock flowed upward altering the shape of the surrounding *bed*s. Earthquakes shoved great blocks of land upward, downward, and sideways. Wind and water eroded the exposed formations. Sediment washed into the ocean where it formed layers. The gradual movement of the land masses or plates sometimes blocked access to open water, and the resulting inland seas slowly evaporated. Great rivers carrying tons of sediment dried up, and became buried by other rocks. Under the right circumstances, these alterations in layers of rock trapped and stored hydrocarbons.

As the earth changed, the weight of overlying rocks continued to push downward, forcing hydrocarbons out of their source rocks. Seeping through subsurface cracks and fissures and oozing through small connections between rock grains, the hydrocarbons began to flow through the porous and permeable rock layers moving in the direction of decreasing pressure. Hydrocarbons generally flowed upward toward the surface until a subsurface barrier stopped them or they reached the earth's surface as an oil seep.

Most of the hydrocarbons did not reach the surface. Instead, they became trapped and were stored in a layer of subsurface rock. Once the hydrocarbons were blocked in a *trap*, they accumulated near the top because hydrocarbons are less dense than the ancient seawater held in the rock layers.

Petroleum Traps

A hydrocarbon reservoir has a distinctive shape, or configuration, that prevents the escape of hydrocarbons that migrate into it. Geologists classify reservoir shapes, or traps, into two types: *structural traps* and *stratigraphic traps*.

Structural Traps

Structural traps form because of a deformation in the rock layer that contains hydrocarbons. Two examples of structural traps are *fault traps* and *anticlinal traps* (fig. 65).

FAULT

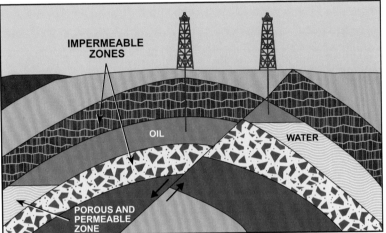

ANTICLINE

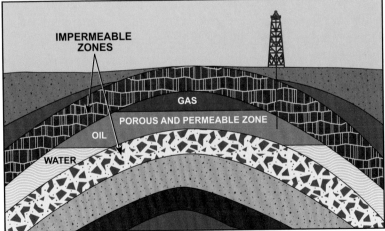

Figure 65.   A fault trap and an anticlinal trap

Fault Traps

A fault is a break in the layers of rock. A fault trap occurs when the formations on either side of the fault move. When petroleum migrates into one of the formations, it becomes trapped there. Often, an impermeable formation on one side of the fault moves opposite a porous and permeable formation on the other side. The petroleum migrates into the porous and permeable formation and cannot get out. If the permeable rock layers do not have an impermeable layer across the fault, the flowing hydrocarbons can move across the fault and are not trapped.

Anticlinal Traps

An *anticline* is an upward fold in the layers of rock, much like a domed arch. The oil and gas migrate into the folded porous and permeable layer and rise to the top because they are lighter than the water also contained in the layer. The oil and gas cannot escape because of an overlying bed of impermeable rock.

Stratigraphic Traps

Stratigraphic traps form when other beds seal a reservoir bed or when the permeability changes within the reservoir bed (fig. 66).

In the drawing labeled A in figure 66, a horizontal, impermeable rock layer cuts off, or *truncates*, an inclined layer of petroleum-bearing rock. Sometimes a petroleum-bearing formation pinches out; that is, an impervious layer cuts it off as shown in the drawing labeled B. Other stratigraphic traps are lens-shaped with impervious layers surrounding the hydrocarbon-bearing rock as shown in the drawing labeled C. Still another type of stratigraphic trap occurs when the porosity and the permeability change within the reservoir. In the drawing labeled D, the upper reaches of the reservoir are impermeable; the lower part is porous and permeable and contains hydrocarbons.

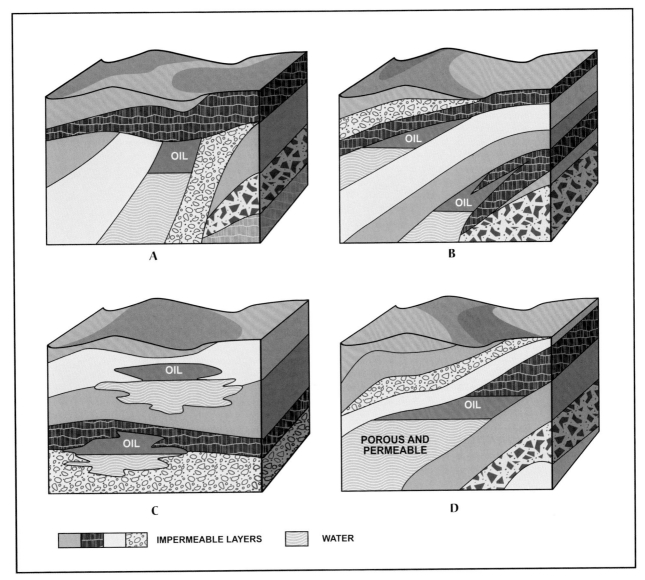

Figure 66.   Types of stratigraphic traps

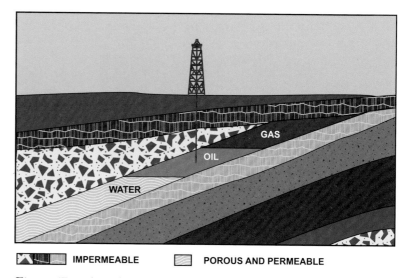

IMPERMEABLE   POROUS AND PERMEABLE

Figure 67.   A combination trap

Other Traps

Many other types of traps can also occur. In a *combination trap*, more than one kind of trap forms a reservoir. An example is a *faulted anticline*. Several faults cut across the anticline. In some places, the faults trap oil and gas (fig. 67).

Another type of trap is a *piercement salt dome* (fig. 68). The famous Spindletop oilwell was drilled into a reservoir formed by a piercement salt dome.

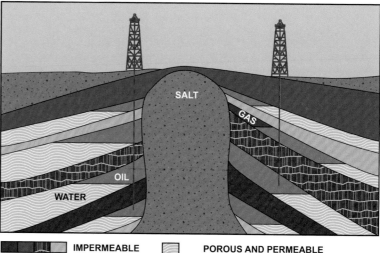

IMPERMEABLE   POROUS AND PERMEABLE

Figure 68.   A piercement salt dome

The salt that forms the dome was originally a solid layer buried beneath other rock layers. Salt is a considered a *plastic* material capable of being moved and shaped. It is less dense than the surrounding rock. The solid, but plastic salt flowed up rock fractures and faults or through unconsolidated deposits. The flow is similar to that of a balloon rising, although at a much slower rate.

Salt domes typically occur in the same geologic region where enough pressure is available to cause the salt to flow and the rock types above the salt are not consolidated enough to prevent the salt from flowing slowly to the surface. Salt domes might be connected to the original layer of salt or they might be a mass of salt that has completely separated from the original layer as it rises toward the surface. Under similar conditions, some *shale* or clay layers form domes called *shale diapirs*.

Finding Petroleum Traps

In the early days of oil exploration, *wildcatters* drilled speculative *wildcat* wells in unproven areas. The term wildcatter originates from the concept of not knowing what to expect, much like when encountering a wild cat.

Wildcatters would attempt to drill in an area close to an oil or natural gas seep. The science of *petroleum geology* developed as wildcatters studied possible reasons to drill wells in particular areas. For example, they looked for features on the surface that indicated subsurface traps. One site was at Spindletop. An underlying salt dome created a hill, or knoll, which seemed out of place on the surrounding coastal prairie. This unusual formation led people like Patillo Higgins and Anthony Lucas to drill for oil based on their previous experience and the study of geology.

Most petroleum deposits are so deeply buried that no surface feature hints at their presence. In many places, such as West Texas, nothing but flat land with minimal features stretches for many miles or kilometres. Yet, the subsurface holds large quantities of oil and gas. Much of the world's oil and gas probably lies offshore, covered by hundreds or thousands of feet (or metres) of water and even more thousands of feet (or metres) of rock.

Figure 69.   To the right of the tire, a large, heavy plate vibrates against the ground to create sound waves.

Over time, scientists have developed effective indirect methods such as *seismology* to view the subsurface. Seismology is the study of sound waves that bounce off buried rock layers. *Geophysicists* create a low-frequency sound on the ground or in the water. The sound could be an explosion or a vibration. Explosion seismology creates sound waves that enter the rock and are recorded. In vibration seismology, a special truck, called a *thumper*, forces a heavy weight against the surface and vibrates the weight (figs. 69 and 70).

Figure 70.   Several special trucks vibrate plates against the ground.

Figure 71.   Fugro Explorer seismic vessel

Because explosions in water can harm marine life, offshore geophysicists use special sound generators (fig. 71).

Regardless of how geophysicists make the low-frequency sound, it penetrates many layers of rock. Where one layer meets another, a boundary exists. Each boundary reflects some of the sound back to the surface. The rest continues downward. On the surface, special devices, termed *geophones*, pick up the reflected sounds, which carry information about the many layers. Cables from the geophones, or *hydrophones* (when monitoring sound in water), transmit the information to sophisticated recording devices in a truck or on a boat (fig. 72).

Figure 72.   Stuck into the ground, a geophone picks up reflected sound waves. Several geophones are placed in an array during a seismic survey.

Geophysicists take the recordings to a laboratory where technicians use computers to analyze and process them (fig. 73). The computers display and print out the *seismic* signals as two- or three-dimensional views.

Some seismic readouts show a cross section of the earth. Others display a top view of buried rock layers. Two-dimensional data can be combined to create a three-dimensional view. This type of display, in effect, removes thousands of feet of rock lying above and around a given layer to reveal the size and shape of the layer. To knowledgeable personnel, seismic displays indicate where oil and gas might exist.

Once an oil and gas reservoir is found, additional *seismic surveys* can show how the reservoir fluids change over time. This is called 4-D seismic because it includes the effect of fluid flow through the reservoir over time. Four-D seismic can show areas in the reservoir that might not have been produced effectively.

Seismic exploration is extremely valuable. Modern seismic technology indicates with increasing accuracy areas that might hold hydrocarbons. However, the only way to be sure that hydrocarbons exist in a given rock layer is to drill a well.

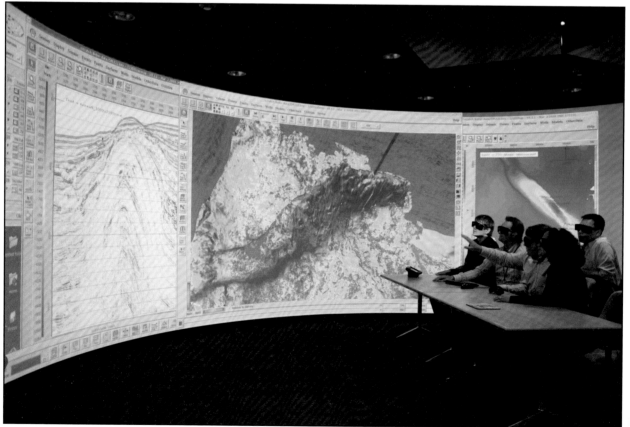

COURTESY OF PHOTOGRAPHIC SERVICES, SHELL INTERNATIONAL LTD.

Figure 73.   iZone Virtual Reality room at EPI Centre in Rijswijk, the Netherlands, 2008. A group of geoscientists looking at data from the Bobo 3D seismic data set, deepwater Nigeria.

TYPES OF WELLS

The petroleum industry generally classifies wells as *exploration wells*, *confirmation wells*, and *development wells*. Other drilling classifications are *infill wells* and *step-out wells*.

An exploration, or wildcat, well is drilled to determine whether oil or gas exist in a subsurface rock formation. The purpose of this type of well is to probe the earth where no known hydrocarbon accumulation exists.

If a wildcat well discovers oil or gas, the company might drill several confirmation wells to verify the well tapped a rock layer with sufficient hydrocarbons for the company to develop it economically.

A development well is one that is drilled in an existing oilfield. A company drills this type of well to remove more hydrocarbons from the field. Engineers carefully study a field's producing characteristics then determine the number of wells required to produce the field efficiently. If a contractor drills between existing wells, they are sometimes called infill wells. If the company drills wells on the edge of an existing field, perhaps to determine the field's boundaries, they might call them step-out wells, or *outpost wells*.

The number of development wells drilled into a particular reservoir depends mainly on its size and characteristics. A reservoir can cover several acres or hundreds of acres (hectares). It might be only a few feet (metres) thick or hundreds of feet (metres) thick. In general, the larger the reservoir, the more wells it takes to produce it. Reservoir characteristics such as porosity and permeability also play key roles. For example, a reservoir with high permeability, which allows the hydrocarbons to flow easily, might not require as many wells to produce as a reservoir with low permeability.

Regardless of the type of well, before the drilling contractor can drill, the operator must prepare the drill site.

The location of the well, or drill site, varies as the surface geography of the earth varies. In the industry's early days, geologists and wildcatters were able to find oil and gas in places readily accessible. As people began using more hydrocarbons, the oil industry extended its search for oil and gas worldwide. Today, companies might drill wells in the frozen wilderness, remote desert, marshes, jungles, rugged mountains, and deep offshore waters. A drill site is anywhere oil and gas exists or might exist.

7
The Drill Site

CHOOSING THE SITE

The operating company considers several factors when deciding where to drill. A key factor is the company knows or believes that hydrocarbons exist in rocks beneath the site. Sometimes, an operator drills a well in an existing field to increase production from it. In other cases, an operator drills a well on a site where no one has previously found oil or gas.

Where no production has occurred, a company often hires geologists and geophysicists to find promising sites (fig. 74). Geologists and geophysicists are called *explorationists* because they explore areas to determine where hydrocarbons might exist. Major companies have an explorationist staff, while independent companies might hire consultants or buy information from companies that specialize in geological and geophysical data.

Figure 74. Geologists working at a prospective petroleum area at the Peel Plateau in the Yukon

Legal and economic factors are also important considerations when selecting a drilling site. The company must obtain the legal right to drill and produce oil or gas on a particular tract. The company must have enough money to buy or lease the right to drill and produce; and the company must have enough money to pay the costs of drilling.

The costs of obtaining a lease and drilling on that lease vary considerably. Costs depend on such factors as the type of reservoir and its depth and location. Offshore and remote sites cost more to drill and produce than readily accessible land sites. A company can easily commit millions of dollars to explore, drill, and produce oil and gas. The rewards can be great but so are the risks and expenses.

An operating company takes several steps before telling the drilling contractor exactly where to place the rig and where to start, or spud, the hole. First, the company carefully reviews and analyzes geologic and seismic records. Legal experts must thoroughly examine lease terms and agreements to ensure the company has clear title and right-of-way to the site. And surveyors establish and verify exact boundaries and locations. The company also confirms it has the necessary funds to finance the project.

On land, operating personnel try to choose a drilling site directly over the reservoir. Ideally, the surface will be accessible and reasonably level. Operating personnel also try to select a location that will not suffer too much damage when the contractor moves in the rig. In an area that is especially environmentally sensitive, the operator and contractor take extra steps to ensure that as little harm as possible occurs to the land. This is frequently done by drilling multiple wells from one surface site.

When drilling offshore, the operator schedules the drilling operation for periods when the weather is expected to be reasonably favorable. If using a bottom-supported rig, a site is selected where the ocean bottom (the *mud line*) can adequately hold the rig supports.

PREPARING THE SITE

At the drilling site, a surveyor marks the exact location of the borehole. On land, this is done simply by driving a stake into the ground. Offshore, the surveyor marks the site with a buoy. On land sites, the operator hires a site-preparation contractor to build a road to access the drilling site and the location to accommodate the rig and auxiliary equipment. If required, bulldozers will clear and level the area. Contractors and operators make every effort to keep damage to a minimum to safeguard the environment. For example, top soil can be stockpiled and placed over the location after the well is finished. Any cleared timber can be stockpiled for sale. In addition to these efforts, the operator is legally required to pay a fee for disruption of the land surface.

Surface Preparation

The contractor uses various materials to prepare the surface and roads around a land location so they can support the heavy weight of drilling equipment. Gravel, or a gravel and clay mixture, might be used. A contractor might place board mats over the road and location to allow access in rainy weather or on land sites where it is too wet to build conventionally. In the far north, permafrost presents a special problem because the heat generated under and near the rig can melt the permafrost, allowing the rig to settle into the thawed soil. In permafrost conditions, the contractor spreads a thick layer of gravel to insulate the area. If gravel is unavailable, polyurethane foam might be used.

Earthen Pits

At a land site, the site-preparation contractor might dig several earthen pits. The earthen pits can be used to hold the active mud volume for drilling the well. More often, the active mud volume is held in steel tanks to make mud conditioning easier and more effective.

A small earthen pit is sometimes used to catch the rock cuttings separated from the active mud volume. This pit is called the *shale pit* because most of the rock cuttings are from shale. Mud and water runoff from the rig and drill site will drain into the shale pit. A larger pit called a *reserve pit* is dug behind the shale pit (fig. 75).

Reserve pits vary in size, depending on how much room is available at the site. Usually, reserve pits are no more than 10 feet (3 metres) deep and are open on top. In the early days of drilling, the reserve pit was mainly a place to store a reserve supply of drilling mud. Today, the waste fluids that collect in the shale pit are pumped or *jetted* into the reserve pit where they are stored temporarily. In an emergency, the fluids held in the reserve pit can be placed back in the active mud system and reused.

Figure 75.   A reserve pit

A small earthen pit might be dug beside the shale and reserve pits to store fresh water for the rig to use to create drilling mud, to wash the rig, or for other uses. A *water well* also might be drilled next to the water pit, or water might be trucked to the site and placed in the water pit. Figure 76 shows a typical layout for earthen pits and their location.

If oil- or salt-water-based drilling fluids are used, the pits will be lined with plastic to prevent soil contamination. To prevent migrating birds from landing in the pits, nets might be placed over the pits. Fences around the pits might be used to keep livestock out.

After the well is completed, the pits are dewatered and all contaminants are removed and disposed of at an approved disposal site. Then the pits are backfilled with a bulldozer and, in some areas, reseeded to prevent erosion. Dewatering a pit can be achieved by evaporating, hauling, or pumping the water to an approved disposal site. Or, the water can be pumped back down the well into a saltwater-bearing formation.

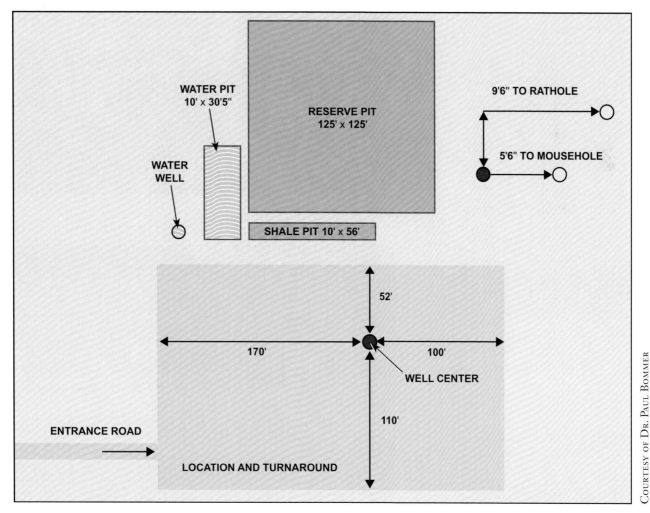

COURTESY OF DR. PAUL BOMMER

Figure 76.   Typical onshore layout of a drilling location

On sensitive land locations, the contractor places drill cuttings into portable receptacles and disposes of them at an approved site (fig. 77a and b). The same method can be used offshore or it might be permissible to discharge small amounts of oil-free and contaminate-free rock cuttings into the water. Most operators and contractors recycle as many drilling mud components and other materials as possible. What they cannot recycle, they discard at approved sites. In some areas, regulatory government agencies or land owners enforce a zero-discharge policy. This policy prevents discarding anything on the ground, into a waterway or estuary, or into the sea. In regulated locations, no earthen pits are used.

Figure 77a.   Pit cleaning with Super Vac units

Figure 77b.   Reserve pit cleanup and removal

Cellars

It is common to dig a *cellar* under a land rig to keep the height of the substructure as low as possible. Cellar sizes vary, but a typical cellar is about 10 feet (3 metres) wide and 10 feet (3 metres) deep. The cellar provides a place to install parts of the wellhead belowground level. If the entire wellhead is built aboveground, the substructure would have to be much taller. Workers line the cellar with boards, corrugated metal, or pour concrete walls to keep it from caving in (fig. 78).

Figure 78.   A concrete pad to support the substructure surrounds this cellar. The conductor casing protrudes from the middle of the cellar. Some of the blowout preventers (BOPs) are shown to the right of the cellar.

Rathole

A special pipe, called the *kelly*, is part of the drill string on some rigs. The kelly rotates the drilling bit and is part of the rig's rotation system. The kelly must be stored when it is not in use. This storage place, called a *rathole*, is a shallow hole drilled off to the side of the main borehole. On an offshore rig, the rathole is a large-diameter length of pipe that extends below the rig floor.

A kelly can be up to 54 feet, or 17 metres, long. Even the tallest land rig substructures are only about 40 feet (12 metres) high, but most are shorter than that. Therefore, the contractor has to drill part of the rathole; otherwise, the rathole would extend too high above the rig floor to be accessible.

On land, the operator might request a special truck-mounted, light-duty unit called a *rathole rig* to drill the rathole. Generally, the rathole rig that drills the rathole is less expensive than the drilling rig. In some situations, after the rig is set up, or *rigged up*, the drilling crew might drill the rathole. In drilled ratholes, the crew extends pipe from the drilled part of the rathole up to the rig floor. The rathole goes through the rig floor and protrudes a few feet, or a half metre or so, above it (fig. 79).

Figure 79.   The kelly has been placed in the rathole when the rig is not drilling. The red kelly drive bushing is sitting on top of the rathole tube with the yellow kelly spinner showing above.

Mousehole

The rathole rig, or the main rig, will also drill a *mousehole* on land sites. A mousehole, like a rathole, is a shallow hole lined with pipe that extends to the rig floor. During drilling operations, the crew puts a length, or joint, of drill pipe temporarily into the mousehole until it is added to the drill string (fig. 80).

As the hole deepens, the crewmembers add the joint to the drill string from the mousehole. A joint of drill pipe is around 30 feet (9 metres) long. If the regular rig's substructure is much shorter than 30 feet in height, the rathole crew also drills a mousehole. For rigs with a top drive, the mousehole is deep enough to hold an entire *stand* of pipe. The stand of pipe might be two, three, or even four joints of drill pipe already *made up* (screwed together). Pipe joints are connected into stands because a top drive can add an entire stand of drill pipe at once while drilling. A rig with a kelly can add only one joint at a time while drilling.

Figure 80.   A joint of drill pipe rests in this rig's mousehole.

Figure 81.   A rathole rig drills the first part of the hole.

Conductor Hole

The rathole rig might also drill the top, or first part, of the main borehole. Occasionally, the operator might save time and money by having the rathole rig unit start, or spud, the main hole before moving in the regular rig. In this process, the rathole crew backs the rathole rig to the cellar (fig. 81).

A special bit starts the main hole in the middle of the cellar. This hole is shallow in depth but large in diameter. Called the *conductor hole*, it might be 36 inches (91 centimetres) or more in diameter. It might be only tens of feet (or metres) deep or hundreds of feet (or metres) deep, depending on surface conditions (fig. 82).

Figure 82.   The conductor hole

The rathole crew lines the conductor hole in the cellar with *conductor pipe* and grouts the conductor pipe in place with cement. Conductor pipe, or casing, keeps the hole from caving in or eroding and enlarging during drilling. It also conducts drilling mud back to the surface when regular drilling begins (fig. 83). Drilling the conductor hole and running the *conductor casing* can also be done by the main drilling rig.

Some shallow soils are soft enough so the conductor pipe can be driven into place with a diesel hammer or *pile driver*. In this case, the conductor pipe is sometimes called *drive pipe*. Offshore where the shallow sediments are soft, the conductor pipe is either driven in using the pile driver or washed in with seawater pumped through a special assembly in the bottom of the pipe. No cement is needed if the conductor casing is driven or washed into place. The contact between the outside of the conductor casing and the sediment is sufficient to form a seal.

Figure 83.   The large-diameter pipe to the right is the top of the conductor pipe. The small-diameter pipe to the left lines the rathole. Later, the rig crew will install a tube to extend the rathole up to the rig floor.

MOVING EQUIPMENT TO THE SITE

After the operator selects and prepares the drill site, the contractor moves the rig to the site. Most land rig equipment is disassembled by crewmembers and moved to the site on trucks. The crew puts the components back together and begins drilling. In remote areas, such as in jungles and Arctic regions, the crew loads the disassembled rig components onto cargo airplanes or helicopters. Boats might tow offshore rigs from one site to another. Some offshore rigs are self-propelled with engines and propellers, providing the power necessary to move it. Sometimes, especially when a rig must be transported long distances, a heavy-lift vessel carries the rig.

Moving Land Rigs

All land drilling rigs are portable. If the rig is small enough to be built on a truck, it is simply driven from one place to another. Once at the site, the rig stays on the truck, the mast is raised hydraulically, and drilling begins (fig. 84).

Rigs too large to fit on one truck are designed differently. Fabricators design medium and large rigs so drilling crews can take them apart, load the components on several trucks, helicopters, or cargo planes, and move them to the drill site. There, crews put the rig together, or *rig-up*. After they drill the well, they dismantle the rig, or *rig-down*.

In deserts and other flat locations, the contractor might *skid the rig*. A rig suitable for skidding has enormous wheels attached to the substructure. When engaged, the wheels allow the rig to be towed short distances without being dismantled. Some onshore rigs drill several wells from one pad, much like an offshore platform. A rig located on a pad that is being moved only a few feet to the next borehole is mounted on rails.

Figure 84.   A portable shallow oil drilling rig

Moving and Setting Up Offshore Rigs

Some offshore rigs are self-propelled. Built-in engines and screws (propellers) move the rig through the water. Rudders similar to those on a ship allow marine personnel to steer the rig while it is moving. Although a self-propelled rig's speed is slow—perhaps 3 or 4 knots per hour, at most—the distances traveled are usually relatively short. To move rigs that are not self-propelled, the contractor hires tow boats to pull them.

The contractor might use a special ship called a heavy lift vessel to carry the rig, whether self-propelled or not, for long moves such as from one ocean to another (fig. 85a and b). Crewmembers position the ship next to the rig for loading. Compartments in the ship are flooded with sea water to submerge the load-carrying deck below the waterline. With the deck below the water surface, large cranes or tug boats pull the rig over the deck of the ship. Pumps remove the water from the compartments and the ship floats back above the water surface with the rig in place on the deck. The process is reversed to unload the rig.

Whether on land or offshore, once the site is prepared for the rig, the next step is for the drilling crew to rig-up; that is, to put the rig components together and prepare the rig for drilling.

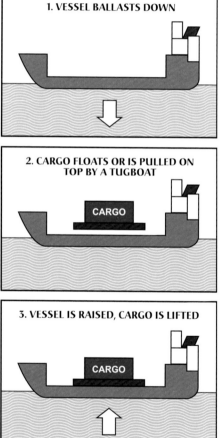

Figure 85a. A heavy lift vessel carries a semisubmersible to a new drilling location.

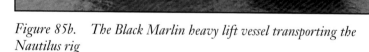

COURTESY OF DOCKWISE, LTD.

Figure 85b. The Black Marlin heavy lift vessel transporting the Nautilus rig

igging up an offshore drilling rig is usually not as complicated
as rigging up a land rig. Most offshore rigs can be moved over
water with almost no need to disassemble major parts. Onsite, the
offshore rig is stabilized by placing rig supports on the ocean floor
for bottom-supported rigs or, by anchors, anchor chains, and wire
or polyester rope for floaters. Only the dynamically positioned
floaters require no additional support to stay in position during
drilling.

To move most land rigs, crewmembers must disassemble many
of its components. Disassembly is required so the parts can be
transported to the next location and then reassembled. For safety,
rigging up usually takes place only during daylight hours. Even with
lighting after dark, there is too much heavy equipment to move
safely during rig-up.

On most land rigs during rigging up, the rig parts are put
back together so the rig can drill a hole. It involves unloading and
hooking up the rig engines, the *mud tanks* and pumps, and other
equipment on the site. One of the last steps, and one of the more
dramatic, is raising the mast from horizontal—the position in which
it was transported—to the vertical drilling position. The first rig
component positioned by the crew is the rig's substructure, which
is the base, or foundation.

8

Rigging Up

SUBSTRUCTURES

A substructure is the framework located directly over the hole; it
is the foundation of the rig. The bottom of the substructure rests
on level ground. The crew places a work platform on top of the
substructure called the rig floor. The substructure raises the rig
floor to approximately 10 to 40 feet (3 to 12 metres) above the
ground. Elevating the rig floor provides room under the rig for
special high-pressure valves and a *blowout preventer (BOP) stack* that
the crew connects to the top of the well's casing. The exact height
of a substructure depends on the space needed for this equipment.
A cellar also provides more space for the equipment.

Figure 86.   A box-on-box substructure

One type of substructure is the *box-on-box* (fig. 86). Using trucks or portable cranes, the crew stacks one steel-frame box on top of another to achieve the desired height.

A more modern substructure is the *self-elevating*, or *slingshot*, type. The crew places it on the site in a folded position (fig. 87). Then the hydraulic pistons are activated to unfold and raise the substructure to full height (fig. 88). Slingshot substructures go up much faster than the box-on-box type.

Regardless of the type, a substructure must be hardy because it supports the weight of the drilling equipment assembled on top and the weight of the entire drill string suspended in the well.

Figure 87.   A slingshot substructure is shown in folded position prior to being raised. The rectangular structures resting on the folded substructure is the rig's doghouse, which is an office and storage space for the drilling crew at the rig floor level.

Figure 88.   The slingshot substructure near its full height

Figure 89.   This drawworks will be installed on the rig floor. Some drawworks are installed below the rig floor.

Figure 90.   The drilling line is spooled onto the drawworks drum.

THE DRAWWORKS

Crewmembers set up many pieces of equipment on the substructure, including a steel-and-wood rig floor on which to work. The *drawworks* is important equipment often supported by the rig floor (fig. 89). In some rig designs, the drawworks might be installed below the rig floor.

Whether at floor level or below, the drawworks is essentially a large hoist. It houses a *spool*, or drum, on which the crew wraps braided-steel cable called *drilling line* (fig. 90). The drilling line is large-diameter wire rope, ranging in size from ⅞ to 2 inches (22.23 to 51 millimetres). It must be big to carry and move huge weight as the well is drilled. The drawworks, the drilling line, the derrick, and the substructure support everything that goes in or comes out of the hole.

RAISING THE MAST OR DERRICK

Crewmembers next raise the mast or derrick. If the rig has a mast, it is raised to a vertical position either by the drawworks or hydraulic pistons (fig. 91). If the rig has a *standard derrick*, the crew bolts it together on the substructure one piece at a time. When the well drilling is done, the crew disassembles the derrick and rebuilds it at a new site.

A standard derrick looks very much like a mast when assembled. A mast is raised or lowered as a complete unit from a pivot point in the substructure. The derrick has four legs that extend from each corner of the substructure (fig. 92). Although different in construction, the term mast and derrick refer to the piece of equipment used to lift long sections of the drill string out of the well. Today, virtually every land rig uses a mast. Offshore rigs use derricks because a derrick does not have to be disassembled after a well is finished.

Crews can raise or lower a mast without completely assembling and disassembling it each time the rig moves. Not having to build and take apart a derrick is a significant time-saving advantage. Once the manufacturer constructs all the braces, girders, and cross members of a mast, no one totally disassembles it again until the contractor scraps it. Some masts fold, telescope, or come apart into sections to make them shorter and easier to move. Nevertheless, they retain their integrity as a unitized structure.

Figure 92. This rig with a standard derrick was photographed in the 1970s at work in West Texas. The rig crew erected it piece by piece, completed the well, took the derrick down the same way, and moved to a new location.

Figure 91. A mast being raised to a vertical position

Derrick and Mast Heights

A derrick or mast supports the entire weight of the drill string and other tools that are *run in* and out of the hole (fig. 93). The drill string might be thousands of feet long. To pull the drill string out of the hole, the crew has to unscrew it, or break it out, into smaller lengths, or stands. To save time, the crewmembers set each stand back in the derrick in the vertical position after they pull it. The tallest derricks are about 200 feet, or 60 metres, high. The shortest are about 65 feet, or 20 metres, high.

Mast or derrick height governs whether the crew pulls pipe from the hole in *singles*, *doubles*, *triples*, or *quadruples*. If crewmembers pull singles, they break out the pipe one joint at a time. A single length, or joint, of drill pipe is about 30 feet, or 9 metres, long. If they pull doubles, they break out the pipe in two-joint stands of about 60 feet, or 18 metres.

Figure 93.   The derrick supports the weight of the drill string and allows the drill string to be raised and lowered.

If they pull triples, they break out the pipe in three-joint stands of about 90 feet, or 27 metres. If they pull quadruples, they break out the pipe in four-joint stands of about 120 feet, or 36 metres. The longer the stand, the faster the crew can pull the pipe and return it to the hole.

Mast Load Ratings

Masts must be strong and, at the same time, portable. Manufacturers rate masts by the vertical load they can carry and by the amount of wind they can withstand. Mast capacities for vertical loads run from 0.25 million up to 1.5 million pounds (113,000 to 680,000 kilograms). Sometimes, the drill string alone might weigh as much as a half million pounds, or over 227,000 kilograms. Most masts can withstand winds of 100 to 130 miles (160 to 210 kilometres) per hour.

RIGGING UP ADDITIONAL EQUIPMENT

Rigging up involves the assembly of a great deal of equipment. Crewmembers install engines and electric generators to power the rig. They set up steel tanks with drilling mud and connect the pumps that will move the mud down the hole. They erect safe stairways and walkways to allow access to the many components of the rig. They also position auxiliary equipment for compressing air, pumping hydraulic fluid, and pumping water.

Truck drivers and *swampers* (their helpers) bring in storage racks, bins, and living quarters for the company representative, the rig superintendent, and the drilling crews. They deliver drill pipe, pipe racks, fuel tanks, wire rope, and other items to the location. Crews place a small metal building called a *doghouse* adjacent to the rig floor (fig. 94). The doghouse is an office for the driller and the crew in which they keep small tools and current drilling records. Each crewmember might be given a locker for clothes and food in the doghouse.

Figure 94. The doghouse is located at the rig floor level.

Within a few hours or days, depending on the size and complexity of the rig, the crew has the rig assembled and is ready to *break tour*; that is, to begin operating 24 hours a day. As mentioned previously, rigging up usually occurs during daylight. When the crews have the rig ready for operation, they begin working three 8-hour tours or two 12-hour tours.

OFFSHORE RIG-UP

Rig-up operations offshore vary with the type of rig. To rig up a platform, for example, a construction crew must build it. After crewmembers pin the jacket to the seafloor, they set up the living quarters and drilling equipment on top of the jacket. Then they add such items as mud tanks, a helicopter deck, and cranes. Much of this equipment is prefabricated on a series of decks called the *topside*.

Mobile offshore rigs require less rig-up time than platforms because most of the equipment is already in place and assembled. Once the crew gets a floating rig anchored or dynamically positioned on the site, drilling operations can begin. Jackups are ready to operate after the crew jacks the legs into contact with the seafloor, raises the drilling deck above the waterline, and scopes out the derrick on cantilevered beams into the drilling position.

Whether the rig is offshore or on land, once crewmembers finish rigging up, they break tour and begin drilling.

The main function of a rotary rig is to drill a hole in the ground, or to *make hole*. Making hole with a rotary rig requires qualified personnel and a large amount of equipment. There are four main categories of equipment systems used in making hole: power, hoisting, rotating, and circulating.

POWER SYSTEM

Every rig needs a source of power to run the hoisting, circulating, and rotating equipment. In the early days of drilling, steam engines powered most rigs. In the 1860s, Colonel Drake powered his rig with a wood-fired steamboat engine. Until the 1940s and 50s, steam engines drove almost every rig (fig. 95).

Steam is a tremendous power source. For example, steam catapults are used today on modern aircraft carriers to launch aircraft. The major problem with using steam power on drilling rigs was that the boilers were heavy and difficult to move. Also, the steam lines to the steam engines were heavy and withstood high pressures and temperatures. Steam power also required large volumes of water and fuel.

9
Rig Components

Figure 95.   In the foreground is a coal-fired boiler that made steam to power the cable-tool rig in the background.

As powerful and portable diesel and gas engines became available, *mechanical rigs* began to replace the steam rigs (fig. 96). Drillers called them mechanical rigs, or *power rigs*, because the engines provided power to the components. In the 1960s and 1970s, electric or *SCR (silicon-controlled-rectification)* rigs used electric motors rather than engines to power the rig. Electricity for the motors was supplied by generators driven by diesel engines. Today, electric or SCR rigs are common, especially for larger rigs.

Figure 96.   A mechanical rig is shown drilling in West Texas in the 1960s. Rigs like this still drill in many parts of the world today.

Figure 97.   Three diesel engines power this rig.

Whether mechanical or electric, virtually every modern drilling rig uses internal-combustion engines as a prime power source, or *prime mover*. Rig engines are similar to those in an automobile, but they are bigger, more powerful, and do not use gasoline as fuel. Most rigs require more than one engine to furnish the necessary power (fig. 97). Rig engines today are primarily diesels because they require less maintenance and have generally superior performance. Diesel fuel has a higher ignition point and, therefore, is safer to transport and store than fuels such as natural gas, LPG, or gasoline.

Depending on its size and the depth of the hole it must drill, a rig might have one to four engines. The bigger the rig, the deeper it can drill, and the more power it needs. Big rigs have three or four engines, producing a combined power up to 3,000 or more horsepower (2,200 kilowatts). Once the engines develop this power, it must be sent or transmitted to other rig components to make them work.

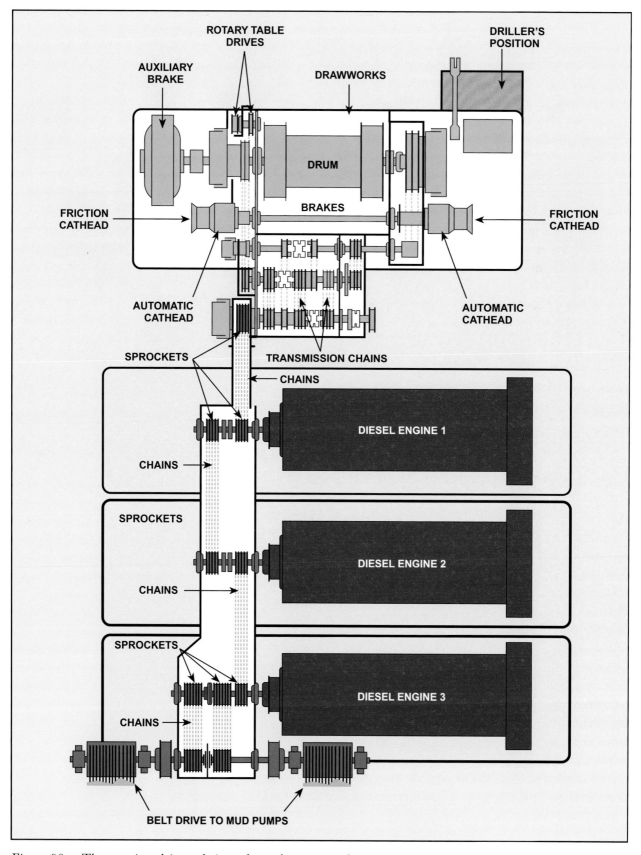

Figure 98.   Three engines drive a chain-and-sprocket compound to power equipment.

Mechanical Power Transmission

A schematic of a mechanical rig is shown in figure 98. It has three 700-horsepower (520-kilowatt) engines hooked up to a *compound*. The compound consists of several heavy-duty sprockets and chains. The sprockets, around which the chains are wrapped, are driven by the engine. The chains drive the various rig components. This chain-and-sprocket arrangement is known as the compound because it compounds or connects the power of several engines.

With compounded engines, the driller can use one, two, or all of them at once. An auxiliary compound used to transmit power to a *mud pump* on the ground is shown in figure 86 in the previous chapter. Although good in theory, this method takes power away from the main rig engines at times when it might be needed. For this reason, auxiliary equipment is not powered by the main rig engines, except on smaller rigs. The auxiliary components, such as the generators and the pumps, are driven directly by independent engines on most mechanical rigs. This allows the engines on the rig to provide power to the hoisting and the *rotating components*.

Electrical Power Transmission

Diesel engines drive large electric generators on electric SCR rigs (fig. 99).

Figure 99.   The diesel engine at right directly drives an alternating current electric generator. This engine-generator set is one of three on this rig.

Figure 100.   Controls in the SCR house where AC electricity is converted to the correct DC voltage for the many DC motors powering this rig.

The generators produce *alternating current (AC)* electricity that flows through cables to a control house called an *SCR house* (fig. 100). In the SCR house, the AC current is converted to *direct current (DC)* and transformed to the correct voltage required by every electric component on the rig.

From the SCR house, electricity goes through more cables to electric motors. DC traction motors of varying horsepower are attached to the equipment to be driven; for example, the drawworks and the mud pumps (figs. 101 and 102).

Figure 101.   A motor-driven drawworks

Figure 102.   Two powerful electric DC traction motors drive the drawworks on this rig.

The SCR system has several advantages over the mechanical system. The electric system eliminates the heavy and complicated machinery that make up the compound. Because an *electric rig* does not require a compound, crewmembers do not spend time lining up and connecting the compound with the engines and drawworks. Motors have a much higher initial *torque* than engines, which improves torque intensive operations such as hoisting and rotating.

On an SCR rig, the engines are placed at a distance from the rig floor making the operation much quieter. An SCR rig could be hooked up to a commercial electric power *transmission* line and run without the diesel generators. This might be done when a rig is located in a city where the engine noise is prohibited or restricted. The major disadvantage of the SCR rig is that skilled electrical technicians are required to maintain the electrical system.

HOISTING SYSTEM

The rig must have a hoisting system to raise and lower the drill string (fig. 103). A typical hoisting system is made up of the drawworks (or hoist), a mast or derrick, the crown block, the *traveling block*, and the wire-rope drilling line.

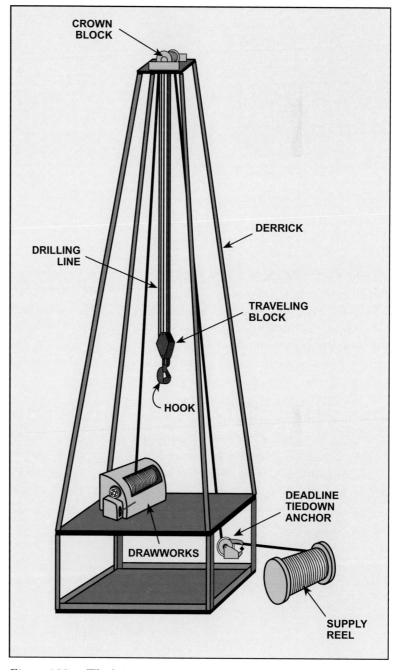

Figure 103.   The hoisting system

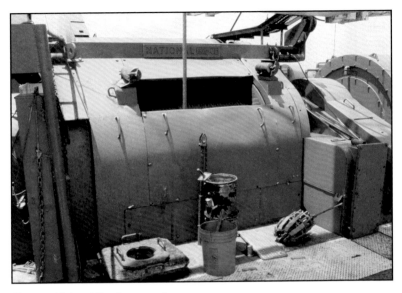

Figure 104.   The drawworks

The Drawworks

The drawworks consists of a revolving drum around which crew-members wrap or spool wire-rope drilling line (fig. 104). It has a *catshaft* on which the *catheads* are mounted. The drawworks also has clutches and chain-and-gear drives so the driller can change the speed and direction of the drawworks.

The driller can slow or stop the drum of the drawworks with the main brake (fig. 105).

Figure 105.   Removing the drawworks housing reveals the main brake bands to the left and right on the hubs of the drawworks drum.

Figure 106.   The electromagnetic brake is mounted on the end of the drawworks. It helps the main brake slow and stop the drawworks drum.

An auxiliary electric brake assists the main brake by absorbing the momentum created by raising and lowering the load (fig. 106). Big electromagnets inside the auxiliary brake oppose the turning forces on the drum and help the main brake stop the load.

Drawworks are rated by the horsepower required to lift the heaviest weight suspended from the derrick at hoisting speed. For example, a drawworks that picks up 600,000 lbs. (272,000 kg) with a block speed of 50 ft/min (15.2 m/min) with an efficiency of 85% requires input horsepower of 1,100 (820 kW).

The Catheads

A cathead is a winch or windlass on which a line, such as rope, cable, or chain, is coiled. When activated, a cathead *reels* in the line with great force. Pulling on a line is vital to making up and breaking out (screwing and unscrewing) drill pipe. Typically, four catheads, two on each end, are mounted on the catshaft of the drawworks.

Note that in figure 98 *friction catheads* are on the ends of the catshaft, and next to them are the *automatic* or *mechanical catheads*. A friction cathead is a steel spool approximately one foot (30 centimetres) in diameter. It revolves as the catshaft revolves. Crewmembers use the friction catheads to move heavy equipment around the rig floor. One floorhand rigs up one end of the *catline* to the object that will be moved. Another floorhand wraps the other end of the catline around the cathead and uses it to lift the object (fig. 107). Releasing or slacking off on the rope allows the cathead to lower the object.

Figure 107.   A floorhand has a fiber rope wrapped around a friction cathead to lift an object on the rig floor. By pulling on the end of the rope, the friction cathead provides the mechanical advantage necessary to lift heavy objects.

Rig personnel also use a small, air-powered hoist to move equipment on and around the rig floor (fig. 108). The air hoist is separate from the drawworks. Air hoists are much easier and faster to use than a friction cathead.

The automatic catheads can be used to make up or break out the drill string when running it into or pulling it from the hole. It is called an automatic cathead because the driller simply moves a control to engage or disengage it. When engaged, an automatic cathead pulls on a chain connected to a pair of tongs to make up or break out the string.

Figure 108.   Floorhand using an air hoist to lift an object

The automatic cathead on the driller's side of the drawworks is the *makeup cathead* because it pulls the tongs clockwise (to the right) when the crew makes up the drill pipe (fig. 109). The automatic cathead on the other side of the drawworks is the *breakout cathead* because it pulls the tongs counterclockwise (to the left) to break out the drill pipe.

On some SCR rigs, the drawworks does not have a catshaft. The friction catheads are eliminated and only the air hoist is used to move objects around the rig floor. The automatic catheads are replaced by independent hydraulic catheads used to pull the tongs.

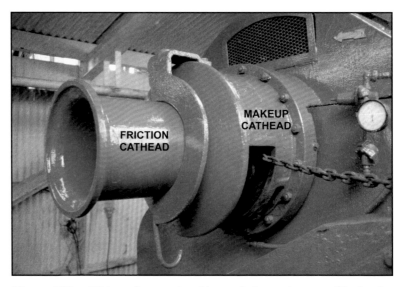

Figure 109.   This makeup cathead has a chain coming out of it that is connected to the tongs. When actuated, the cathead pulls on the chain to apply tightening force to the tongs.

The Blocks and Drilling Line

Drilling line runs from ⅞ to 2 inches (22 to 51 millimetres) in diameter, depending on how much load it must pick up. Wire rope is similar to fiber rope, but it is made out of steel wires (fig. 110). It looks much like common cable but is designed especially for the extremely heavy loads encountered on the rig.

The line comes off a large reel, or a *supply reel* (fig. 111). From the supply reel, it goes to a strong clamp called the *deadline anchor* (fig. 112).

Figure 110. Wire-rope drilling line coming off the drawworks drum

DRILLING LINE

Figure 111. Drilling line is stored on this supply reel at the rig. When needed, the line can be taken off the reel to replace worn line.

Figure 112. Drilling line is firmly clamped to this deadline anchor.

SHEAVES

Figure 113.   The sheaves (pulleys) of this crown block are near the bottom of the photo.

From the deadline anchor, the drilling line runs up to the top of the mast or derrick to a set of large *pulleys*. In the oilfield, the pulleys are called *sheaves* (fig. 113). The completed set of sheaves is called the crown block.

The drilling line is strung or threaded several times between the crown block and another large set of sheaves called the traveling block. Because the line is strung several times between the crown block sheaves and the traveling block sheaves, the hoisting effect is equivalent to several lines. For example, a weight picked up by blocks strung on eight lines requires a force on the *fastline* at the drawworks of ⅛th of the weight that is picked up if friction is ignored. The fastline is the end of the drilling line connected to the drum or reel of the drawworks. The heavier the anticipated loads on the traveling block, the more times the line is strung between the crown and traveling block. In figure 114, ten lines are strung, which means the line was strung five times between the crown and traveling block. The hoisting speed slows as more lines are strung between the blocks. Therefore, no more lines are strung than those needed to speed up hoisting operations.

Once the last line is strung over the crown block sheaves, the end of the line goes down to the drawworks drum where it is firmly clamped. Then the driller takes several wraps of line around the drum (fig. 115).

Figure 114.   Ten lines are strung between the traveling block and the crown block.

Figure 115.   Several wraps of drilling line on the drawworks drum

Running from the drawworks to the crown block, the fastline moves as the driller raises or lowers the traveling block in the mast or derrick. The end of the line that runs from the crown block to the dead line anchor is the *deadline;* the term dead refers to no movement.

The number of sheaves on the crown block is always one more than the number of sheaves on the traveling block. For example, a ten-line string requires six sheaves in the crown block and five in the traveling block. The extra sheave in the crown is needed for stringing the deadline.

Attachments to the traveling block include a spring that acts as a shock absorber and a large *hook* from which crewmembers suspend the drill string (fig. 116). Also, hanging from the traveling block is an *elevator.* The elevator is the clamp latched below a *tool joint* on the drill pipe used to raise or lower the drill string during *trips.* On floating rigs, a heave compensator (see figure 45b) is also attached to the traveling block so the drill string does not change position when the vessel moves up and down with the waves.

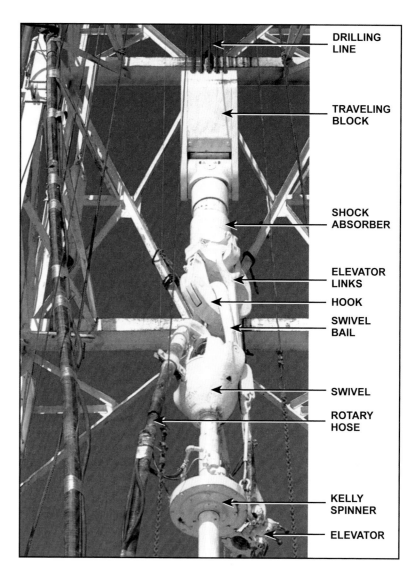

Figure 116.   Traveling block and kelly assembly

Masts and Derricks

Masts and derricks are tall structural towers that support the blocks and drilling tools (fig. 117). They also provide height to allow the driller to raise the drill string so crewmembers can break it out and make it up. A mast is a portable derrick that crewmembers can raise and lower as a unit. A standard derrick, on the other hand, requires that crewmembers assemble and disassemble it piece by piece. They cannot erect it or take it down as a single unit. Most land rigs today use masts because they rig up and down much quicker than standard derricks.

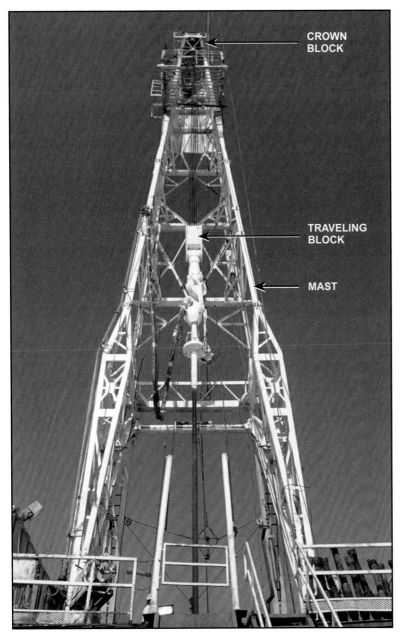

Figure 117.   The mast supports the blocks and other drilling tools.

Manufacturers must make masts strong and portable for onshore use. The loads a mast might support on a deep well can be as high as 2 million pounds, or 1,000 tons (907 tonnes). After finishing up one hole, crewmembers usually move the rig several miles to begin another. Some of the specifications manufacturers use to rate masts and derricks are their height, the vertical load they can carry, and the wind load they can withstand from the side. For example, a mast might be 136 feet (41.5 metres) tall, able to support 275 tons (249 tonnes), and capable of withstanding 100-mile-per-hour (161 kilometres per hour) winds. These are impressive specifications for such strong equipment that is relatively easy to move.

The front of the derrick always has a space that allows pipe and tools to be hoisted up from the ground and suspended in the derrick. The open front of the derrick is called the *V-door*. It is so named because the open space between the legs of the original standard derrick had a v-shape. In front of the V-door is a ramp that goes down to a platform just above the ground called the *catwalk*. The catwalk is a raised level platform where pipe can be picked up or new tools made ready for raising up into the derrick. The ramp that connects the catwalk to the V-door is called the V-door ramp. Figure 92 shows a typical inverted V-door, V-door ramp, and catwalk.

ROTATING SYSTEMS

Rotating equipment is the mechanism that turns the bit. In general, rigs can rotate the bit using equipment in one of three ways. The traditional system uses a rotary table and kelly. A second method uses a *top-drive system*. A third system uses a *downhole motor*. The downhole motor can be used with or without the rotary table or a top drive.

Rotary-Table System

A *rotary-table system* consists of five main parts: (1) a rotary table with a turntable, (2) a *master bushing*, (3) a *kelly drive bushing*, (4) a kelly, and (5) a *swivel* (fig. 118).

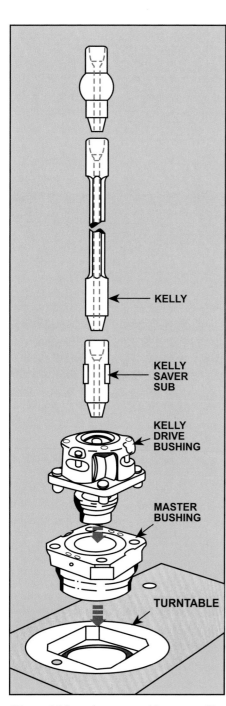

KELLY

KELLY SAVER SUB

KELLY DRIVE BUSHING

MASTER BUSHING

TURNTABLE

Figure 118.   A rotary-table system (See figure 116 for the location of the swivel.)

Turntable

The rotating turntable is housed in a stationary heavy-duty rectangular steel case (fig. 119). The turntable is round and located near the middle of the case. It produces a turning motion that is transferred to the pipe and bit. The case also holds gears and bearings on which the turntable rotates. The turntable is powered by an electric motor or gears and chains from the transmission of a mechanical rig.

Figure 119.   *The turntable is housed in a steel case.*

Master Bushing and Kelly Drive Bushing

A *bushing* is a fitting that goes inside an opening in a machine. A rotary table master bushing fits inside the turntable (fig. 120). The turntable rotates the master bushing. The master bushing has an opening through which crewmembers *run pipe* into the *wellbore*.

Figure 120.   The master bushing fits inside the turntable. A tapered bowl fits in the master bushing.

TAPERED
BOWL

Figure 121.   Crewmembers are installing one of two halves that make up the tapered bowl.

A *tapered bowl* fits inside the master bushing (fig. 121). This bowl serves a vital function when the pipe and bit are not rotating. When the driller stops the rotary table and uses the rig's hoisting system to lift the pipe and bit off the bottom of the hole, it is often necessary for crewmembers to suspend the pipe off the bottom of the hole. To hold the pipe in place, they put a set of segmented pipe-gripping equipment called *slips* around the pipe and into the master bushing's tapered bowl (fig. 122). The slips firmly grip the pipe to keep it suspended off the bottom.

The kelly drive bushing is the third piece of rotary equipment. It transfers the master bushing's rotation to a special length of pipe called the kelly. The kelly drive bushing fits into the master bushing. There are two main types of master and kelly drive bushing. One master bushing has four drive holes (fig. 123). Strong steel *pins* on the bottom of a kelly drive bushing, made for this type of master bushing, fit into the holes. When the master bushing rotates, the pins engaged in the drive holes rotate the kelly drive bushing.

Figure 123.   The master bushing has four drive holes into which steel pins fit on the kelly drive bushing.

Figure 122.   Crewmembers set slips around the drill pipe and inside the master bushing's tapered bowl to suspend the pipe.

Another type of master bushing has a square opening and no drive holes. The opening corresponds to a square shape on the bottom of a kelly drive bushing made for this type of master bushing. A square kelly drive bushing does not have drive pins; instead, the square bottom of the kelly drive bushing fits into the corresponding square opening in the master bushing (fig. 124). With the square drive bushing in place, the rotating master bushing turns it.

Figure 124.   A master bushing with a square bottom that fits into a square opening in the master bushing

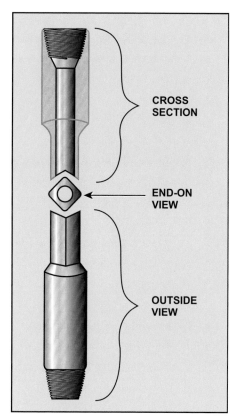

Figure 125a.   A square kelly

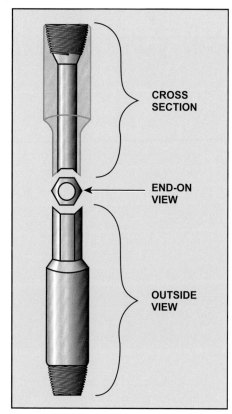

Figure 125b.   A hexagonal kelly

Kelly

The fourth major component of a rotary-table system is the kelly. A kelly is a pipe with an unconventional shape. It is not round; instead, it has four or six flattened sides that run almost its entire length (figs. 125a and 125b). Kellys are square or *hexagonal* in cross section, because the flat sides enable rotation of the kelly. The kelly's flat sides fit into the corresponding square or hexagonal opening in the kelly drive bushing.

When the driller inserts the kelly into the corresponding kelly drive bushing and activates the rotary table, the kelly drive bushing rotates the kelly (fig. 126). The drill string is attached below the kelly and turns whenever the rotary table turns the kelly.

As it rotates, the kelly can move freely up or down through the bushing opening. The kelly is gradually lowered as the bit drills deeper. When the driller stops rotating and lifts (picks up) the drill string, the kelly slides through the kelly drive bushing as the hoisting system raises the kelly. When the drive bushing reaches the bottom of the kelly, the kelly tool joint, which is bigger than the bushing's opening, keeps the bushing from sliding off the kelly. At this point, if the drill string hoisting continues, the kelly drive bushing is raised out of the rotary table master bushing.

Figure 126.   A hexagonal kelly inside a matching opening in the top of the kelly drive bushing

A hexagonal kelly is generally stronger than a square kelly. Consequently, contractors tend to use hexagonal kellys on large rigs to drill deep wells because of their extra strength. Small rigs often use square kellys because they are less expensive. Manufacturers make kellys according to *American Petroleum Institute (API)* specifications. The API is a trade association that sets oilfield standards and specifications. A standard API kelly, either square or hexagonal, is 40 feet (12.2 metres) long, although an optional length of 54 feet (16.5 metres) is available. The length of the kelly determines how much new hole can be drilled before adding another joint of drill pipe to the drill string. Drill pipe commonly comes in 30-foot (9.2-metre) lengths, so the 40-foot kelly is the most widely used.

Swivel

The fifth principal part of a rotary-table system is the swivel. The swivel allows the kelly and drill string to rotate without twisting the drilling line while suspended by the blocks. A heavy-duty bail, similar to but much larger than the handle on a water bucket, fits into a big hook on the bottom of the traveling block (fig. 127). The hook suspends the swivel and attached drill string.

The crew makes up the top of the kelly to the swivel. The kelly screws onto a threaded fitting, or *stem*, that comes out of the swivel. This stem rotates with the kelly, the drill string, and the bit. At the same time, drilling mud flows through the stem and into the kelly and drill string.

Figure 127.   The hook on the bottom of the traveling block is about to be latched onto the bail of the swivel.

Near the top and on one side of the swivel is a *gooseneck* (fig. 128). The gooseneck is a curved, erosion-resistant piece of pipe that conducts drilling mud under high pressure into the *swivel stem*. A special hose—the rotary, or kelly, hose—attaches to the gooseneck. The *rotary hose* conducts drilling mud from the pump to the swivel, and because the hose is flexible, it allows the kelly to be raised or lowered while circulating.

To summarize, the kelly-and-rotary-table system includes:

- The turntable in the rotary table that rotates the master bushing
- The master bushing that rotates the kelly drive bushing
- The kelly drive bushing that rotates the kelly
- The kelly that rotates the attached pipe and bit

Figure 128. Drilling fluid goes through the rotary hose and enters the swivel through the gooseneck.

- The swivel that suspends the pipe, allows it to rotate, and enables passage of the drilling mud into the kelly and drill string

Top Drive

The use of a top drive replaces the kelly, the kelly drive bushing, and the rotating master bushing. The top drive, also called a *power swivel*, rotates the drill string and bit (fig. 129). Like a regular swivel, a top drive hangs from the rig's large hook and has a passageway for drilling mud to flow into the drill pipe. The top drive also has a set of elevators for use during trips and is equipped with an electric or hydraulic heavy-duty motor. Some top drives have two motors. Drillers operate the top drive from a control console on the rig floor.

The motor turns a threaded drive shaft. The crew *stabs* (inserts) the drive shaft into the top of the drill string. When the driller starts the top drive's motor, it rotates the drill string and bit. Rigs with a top drive still need a rotary table with a master bushing and bowl to provide a place to suspend the pipe on slips when the bit is not drilling.

Figure 129.   A top drive, or power swivel, hangs from the traveling block and hook.

There are several advantages to a top drive. The top drive suspends one stand (typically 3 joints) of drill pipe and uses this entire stand to drill new hole. When using a kelly, only 40 feet of new hole can be drilled before adding one joint of pipe. Therefore, the top drive is faster because it saves connection time.

When drilling a directional hole, orienting the direction of a drill bit with a top drive can be accomplished faster than when using a kelly. A top drive can rotate in either direction allowing the drill string to back ream parts of the hole if necessary.

Top drives are mounted on a track that runs vertically in the derrick to ensure the top drive remains centered. Because the top drive blocks the driller's view of the top of the traveling block, a camera is mounted in the derrick so the driller can see the traveling block. The camera helps the driller prevent the traveling block from accidentally colliding with the crown block at the top of the derrick.

Downhole Motors

In certain situations, the rig might use a downhole motor to rotate the bit. Unlike a rotary-table or a top-drive system, a downhole motor does not rotate the drill pipe. The downhole motor rotates only the bit. The motor is installed just above the bit in the drill string.

Pressure from the drilling mud powers the downhole motor (sometimes called a *mud motor*). Drilling mud is pumped down the drill string, and when the mud enters the motor, it strikes a spiral shaft inside the motor housing. The shaft (called the *rotor*) and housing (called the *stator*) are offsetting helical shapes so the mud pressure will cause the rotor to turn. As shown in figure 130, the rotor will turn with an off-center eccentric motion. For this motion to be converted to a circular motion at the bit, the rotor is attached to a titanium shaft that will flex but not break. The bit is attached to the flexible shaft and the shaft turns the bit.

As shown previously in figure 21, there is a thrust bearing on the bottom of the motor that allows the bit to continue turning when drill string weight is applied.

A downhole motor can rotate faster than a top drive or a rotary table, reaching speeds of 200 rpm or more. When high bit speed is desired, a motor can be used. The rotation of the drill string is also desirable to minimize friction in the hole. Therefore, even when using a downhole motor, the top drive or rotary table will often be turned as well. The rotating speed of the surface equipment is added to the speed of the motor when turning the bit.

Rigs often use downhole motors to drill directional holes. A *directional hole* is a hole intentionally drilled to deviate from vertical.

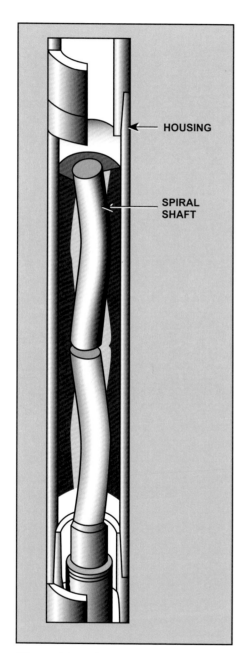

Figure 130. Mud pressure pumped through the drill string forces the spiral rotor of the mud motor to turn inside the rubber helical-shaped stator. The drill bit is connected to the rotor through a flexible titanium shaft and bearing so the bit will turn when the rotor turns.

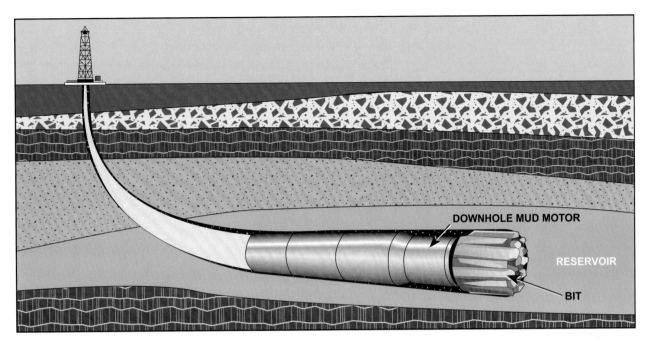

Figure 131.   Horizontal hole

It is sometimes necessary to drill a hole on a slant because a vertical hole will not reach the required part of a reservoir, or because a vertical well does not allow the most efficient recovery from the reservoir (fig. 131).

To steer the bit in the proper direction, a special mud motor with a bend in the motor housing is used (figs. 132 and 133). The small bend in the motor can be oriented to steer the bit in the desired direction. During steering operations, only the motor is used to turn the bit so that the bend in the motor always points the bit in the correct direction.

Figure 133.   An adjustable bent housing on the motor deflects the bit a few degrees off-vertical to start the directional hole.

Figure 132.   A downhole motor lying on the rack prior to being run into the hole

The Drill String

The drill string can consist of many different tools. The simplest drill string consists of drill pipe and heavy-walled pipe called *drill collars* (fig. 134). Drill collars, like drill pipe, are steel tubes through which the drilling mud is pumped.

Figure 134.   Drill collars are placed on the pipe rack prior to being run in the hole. Note the box end threads are covered by plastic protectors.

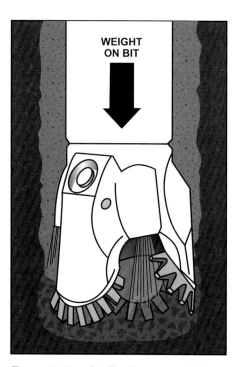

Figure 135.   Drill collars put weight on the bit, which forces the bit cutters into the formation to drill it.

Drill collars are thicker than drill pipe and weigh more on a pound-per-foot (kg/metre) basis than drill pipe. Drill collars are used in the bottom part of the string and must be heavy. Most of the drill string is made up of drill pipe, but enough drill collars must be used to exert sufficient weight on the bit. As the weight of the drill collar presses down on the bit, the bit cutters bite into the formation and drill it (fig. 135). The drill collars, any other tools placed below the drill pipe, and the bit are called the *bottomhole assembly (BHA)*.

Drill collars are roughly 30 feet (9.1 metres) long. Those made to API specifications range in diameter from 2⅞ to 12 inches (73.03 to 304.8 millimetres). One type of drill collar is 30 feet (9.1 metres) long, 6.25 inches (158.75 millimetres) in outer diameter, 2.875 inches (73 millimetres) in inner diameter, and weighs 2,470 pounds (1,120 kilograms). If the BHA of this particular drill collar consisted of twelve joints, the assembly would have an air weight of 29,640 pounds (13,444 kilograms).

A length or joint of drill pipe is about 30 feet (9.1 metres) long (fig. 136). Each end of each joint is threaded. One end has inside, or female threads, and is called the *box*; the other end has outside, or male threads, and is called the pin.

When crewmembers make up drill pipe, they insert, or stab, the pin end into the box and tighten the *connection* (fig. 137). They call the threaded ends of drill pipe tool joints. The tool joints are welded onto the ends of the drill pipe and provide a high-strength, leak-proof connection that can withstand, not only the large *tensile forces* created by the long length of the drill string, but the high torque created when the drill string rotates.

Figure 136.   Several joints of drill pipe are placed on the pipe rack before being run in the well. Note the pin ends of the tool joints are covered with metal thread protectors.

Figure 137.   A floorhand stabs the pin of a joint of drill pipe into the box of another joint.

Figure 138. Two drill collars on a pipe rack; at left is the drill collar box; at right is the pin

Tool joints are not added to drill collars because the walls of drill collars are thick and the threads can be cut directly into and on the drill collar ends. Like drill pipe, drill collars also have a box and pin end (fig. 138).

Drill pipe can be distinguished from drill collars because drill collars do not have the bulge at either end that characterizes the tool joints of drill pipe (fig. 139).

Figure 139. Drill collars racked in front of drill pipe on the rig floor

Bits

The drill bit is the cutting structure that mechanically breaks the rock at the bottom of the hole. Bits are made in various types and sizes, depending on the job. They are designed to drill a specific size hole in a particular kind of formation.

Bits fall into two main categories: (1) *roller cone* and (2) *drag bits*. Both have cutters, which break rock as the bit drills. Bits have several kinds of cutters. The cutters for roller cone bits are either steel teeth or *tungsten carbide inserts*. Cutters for drag bits are steel blades, natural diamonds, or synthetic polycrystalline diamond compacts (PDC).

Roller Cone Bits

Roller cone bits have steel cones that roll, or turn, as the bit rotates (fig. 140). The cutters are on the end of the cones. As the cones roll over the bottom of the hole, the cutters crush and shear the rock into relatively small chips or cuttings. Drilling fluid, which comes out of nozzles in the bit, removes the cuttings. Roller cone bits have three cones. The cutters on a roller cone bit are either steel teeth or tungsten carbide inserts. Manufacturers *mill* (cut) or *forge* (hammer) the teeth out of the steel body of the cones.

For tungsten carbide insert bits, holes are drilled in the cones and the tungsten carbide cutters are press-fit into the holes (fig. 141). A long tooth or long insert is used to drill soft formations. The harder the formation, the shorter the tooth or the insert becomes so the bit cutter is not immediately broken against the hard rock.

Figure 140.   A roller cone bit has teeth (cutters) that roll, or turn, as the bit rotates.

Figure 141.   Tungsten carbide inserts are tightly pressed into holes drilled into the bit cones.

Roller cone bits have one jet nozzle for each cone and sometimes have one nozzle in the center of the bit. The nozzles are pointed in such a way that they help keep the cutters clean. The nozzles in the bit can be changed to different sizes. The nozzle sizes can be selected by engineers to help optimize the cutting speed of the bit. As a cone moves, a fluid stream flows from the drill string through nozzles on the bit. The fluid hits the rock, chips, and lifts the cuttings from the bottom of the hole. If the cuttings are not removed from the area, the cutters will contact them again, lowering the cutting efficiency and slowing the *rate of penetration (ROP)* through the formation (fig. 142).

Figure 142.   Drilling fluid (salt water in this photo) is ejected out of the nozzles of a roller cone bit.

Inside each roller cone, the bits have bearings that allow the cone to turn as the drill collar weight is applied (fig. 143).

The bearings are categorized as unsealed, sealed, or journal types. The unsealed bearing has the shortest rotating life, and the journal type has the longest life. Bearings are often the limiting factor for roller cone bits. When the end of bearing life is reached, the cone could stop turning and can come off the bit. If this occurs, the cone is left behind at the bottom of the well and must be retrieved before more drilling can be done. The drawback of roller cone bits is their moving parts (the cones) have a predetermined number of hours of bearing life regardless of whether the cutters are dulled or not.

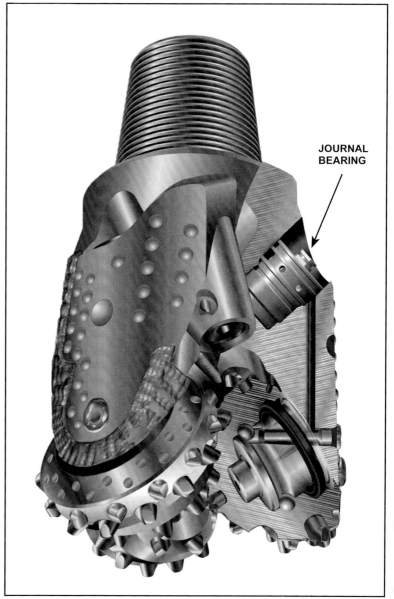

JOURNAL
BEARING

COURTESY OF HALLIBURTON

Figure 143. Bit cutaway showing internal bearing

Drag Bits

Drag bits have no cones and no moving parts. The cutters are inserted into the shape or *matrix* of the bit. The bit turns against the bottom of the hole when the drill string and/or motor turn and the cutters shear the rock in contact with the bit. The term drag bit is used because the bit drags across the bottom of the hole and does not roll, because it has no bearings and cones. This is a major advantage over roller cone bits because there is no bearing life limitation. A drag bit will continue to drill as long as the cutters are intact.

The cutters are longer or larger for soft formations and shorter or smaller for hard formations. The position of the cutters on the matrix will be different depending on the type of rock to be drilled. For extremely hard formations, industrial-grade natural diamonds are placed in the matrix (fig. 144). Diamonds are the hardest type of cutters available and are very small. The natural *diamond bits* last a long time when drilling the hardest types of formations.

For almost every other type of formation, a synthetic *polycrystalline diamond compact (PDC)* can be used (fig. 145). The PDC can be made any size and placed in any pattern to fit a particular type of formation.

A drag bit can also be made with steel blades. Normally, the drag bit is used only to clean out inside wells when removing drilling cement or washing out sand. The steel blades used for drilling formations would not stay sharp for long. The drag bits can have many jet nozzles, and the nozzles perform the same service as those for roller cone bits. The nozzles are designed to help keep the cutters clean and optimize cutting speed. PDC drag bits generally last longer and drill faster than roller cone bits. However, PDC bits are considerably more expensive than roller cone bits.

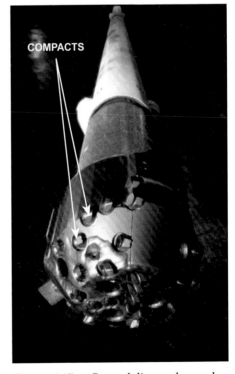

Figure 145.   Several diamond-coated tungsten carbide disks (compacts) form the cutters on this polycrystalline diamond compact (PDC) bit.

Figure 144.   Several types of natural diamond bits are available.

Weight on Bit and Rotating Speeds

Putting weight on a bit makes its cutters bite into the rock. Drillers usually apply weight on the bit by allowing some of the weight of the drill collars above the bit to press down on it. The amount of weight used depends on the size and type of bit and the speed at which the driller rotates it. The amount of weight also depends on the type of formation being drilled. Roller cone bits require more bit weight and less rotating speed, because the cutters on a roller cone bit must first crush the rock. The PDC drag bits operate best at high rotating speeds and less bit-weight because the cutters shear the rock rather than initially crushing it.

CIRCULATING SYSTEM

The rotary rig circulation system moves drilling fluid down the inside of the drill string, out the bit nozzles, and back to the surface equipment in the *annular space* between the outside of the drill string and the hole. The important role of drilling fluid along with the capability of circulating drilling fluid makes rotary drilling the method of choice in the industry.

Drilling Fluid

Drilling fluid can be divided into two broad categories: liquid and gas. Liquid drilling fluid is often called drilling mud. Both categories of drilling fluid must clean rock cuttings from the hole and cool and lubricate the bit and drill string.

Drilling mud stabilizes the borehole to prevent cave-in and collapse. By creating pressure inside the hole that is greater than the reservoir pressure, drilling mud prevents the flow of reservoir fluids into the borehole. This is done by adjusting the density (*mud weight*) of the liquid. The mud is stored in tanks where it can be treated (fig. 146).

Figure 146.   Drilling mud swirls in one of several steel tanks on this rig.

In the tanks, the weight of the mud can be increased by adding heavy minerals such as barium sulfate (*barite*). Adding barite can make mud weight as high as 19 pounds per gallon (ppg) or 2.3 kg/l. A mud balance is used to measure the density of the mud weight (fig. 147).

The base liquid in drilling mud can be water or oil. If water is used, it can be fresh water or seawater. The water is thickened (higher viscosity) by adding clay and polymers, which help suspend the cuttings. The clay forms a wall cake against the permeable portions of the hole to help prevent collapse and limit mud seepage into the formations.

Oil-base drilling mud might be diesel oil or synthetic oil. The oil is emulsified with water to provide the same properties as the water-base mud. The water in a water-base mud causes some shale to swell and heave into the borehole. Correctly made *oil-base* mud prevents this problem. Also, oil-base mud provides better lubricating properties and often allows the bit to drill faster than a water-base mud. Synthetic oil-base mud is not an environmental hazard like diesel-base mud. The drawbacks to oil-base mud are:

- It is more difficult to handle at the rig.
- It can be an environmental hazard if it is diesel oil-based.
- It is considerably more expensive than water-base mud.

Gas drilling fluids can be air, air and water foam, or natural gas. When drilling with gas, the reservoir pressures cannot be contained by the drilling fluid pressure in the borehole. This is because, at the pressures used in drilling, gas has a low density compared to liquid mud. Also, gas does not provide significant hole-stabilization and lubricating properties of liquid mud. Consequently, gas drilling is confined to areas where the rock is stable and will not cave in, and where significant flows of reservoir liquids (water or oil) will not occur during drilling.

Figure 147. A derrickman measures the density (weight) of a drilling mud sample using a balance calibrated in pounds per gallon.

In areas where gas drilling can be used, the advantage of gas is the pressure inside the borehole is kept low and greatly speeds the drilling process. Gas does not require significant additions of chemicals, clay, barite, or liquid, making it relatively inexpensive to create. Mud mixing and cutting separation equipment are not used. However, to divert the returning gas flow away from the rig, *wellhead* equipment called a *rotating head* is required. The return flow exits the well under the rotating head and travels down a large-diameter pipe called a *blooie line*. The end of the blooie line is placed in the large reserve pit where the gas is vented, or *flared*, and the cuttings are allowed to fall into the pit. Several gas compressors, normally rented for the job, circulate the gas. Special drill bits that can withstand higher operating temperatures caused by the friction associated with gas drilling might be required.

Today, expensive natural gas is not often used for gas drilling because it must be flared at the end of the blooie line. For environmental reasons, flaring is no longer permitted except as an emergency procedure. Regardless of the type of drilling fluid used, the derrickman is responsible for maintaining the correct fluid properties as directed by the mud engineer, toolpusher, and company representative.

Circulating Equipment

The circulating equipment for gas drilling fluids consists of the rotating head and blooie line to divert the returning gas flow and compressors and create sufficient pressure in the gas to make it flow. However, the equipment required for liquid mud is more extensive.

Clean mud is pumped from the suction tank through large duplex or triplex mud pumps. There are normally two and sometimes three mud pumps to supply the flow rate and pump pressure desired by the *drilling engineer* (fig. 148).

To increase the total flow rate, the pumps can be operated in parallel at the same pressure or operated as single units with the other pumps on standby. As shown in figure 149, the pump pressure is enough to cause the mud to circulate from the pump discharge through the circuit and return to *shale shaker*.

Figure 148.   Powerful mud pumps (most rigs have at least two) move drilling mud through the circulating system.

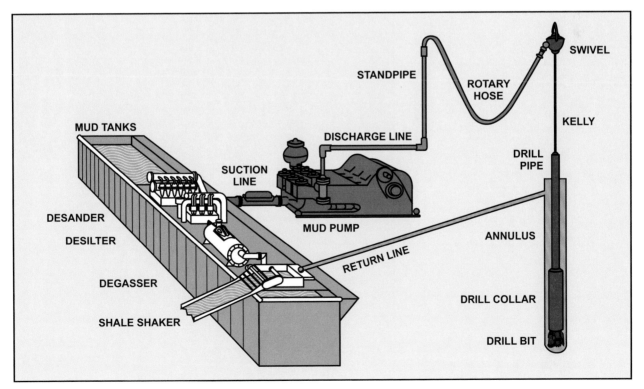

Figure 149.   Components of a rig circulating system

The mud is discharged under pressure from the pumps. It flows through the surface lines, up the *standpipe* welded to a derrick leg, and through the rotary hose to the top drive or swivel. The standpipe and the rotary hose allow mud circulation to continue while the drill string is raised or lowered in the derrick (fig. 150). The mud flows through the inside of the drill string and exits the nozzles of the bit.

Once the mud leaves the bit, it picks up cuttings and flows back to the surface. At the surface, the mud flows by gravity down the return line into the first in a series of machines that remove the drilled solids and any gas from the mud. The first solids-removal machine is a vibrating screen called a shale shaker (fig. 151). The shale shaker is appropriately named because it rapidly vibrates, or shakes, as the mud falls through the screen, sifting out the largest cuttings. Except in environmentally sensitive areas on land, the cuttings fall into an earthen shale pit excavated during site preparation.

Figure 150. The standpipe runs up one leg of the derrick, or mast, and conducts mud from the pump to the rotary hose.

Figure 151.   Mud with cuttings falls over the vibrating shale shaker screen. The mud and fine particles fall through the screen into the first mud tank. The cuttings too large to pass through the screen are vibrated off the end of the shaker and into the shale pit.

In environmentally sensitive areas where the contractor cannot use earthen pits, the *shaker* dumps the cuttings into a steel tank for future disposal in an approved area. Offshore, cuttings are usually emptied into a barge and transported for proper disposal at an approved land site. If the cuttings are oil-free, it might be permissible to discharge them into the water. If more water must be added to the mud, it is done as the mud flows out of the shale shaker.

The next solid-removal machines are the *desander* and *desilter* (figs. 152 and 153). Both machines are cone-shaped *hydrocyclones* that spin the heavier solids to the outside of the core by centrifugal force and discard them out the bottom of the cone and into the shale pit.

Figure 152.   Desanders remove sand-sized particles from the mud.

Figure 153.   Desilters remove smaller silt-sized particles from the mud.

Figure 154.   The degasser removes a relatively small volume of gas that enters the mud from a downhole formation and is circulated to the surface in the annulus.

The solids-free mud flows out the top of the cones and back into the tank. The mud must be free of gas for the desander and desilter to work properly, so a *vacuum degasser* is placed in front of them (fig. 154). The *degasser* can remove small amounts of gas from mud when a small vacuum is placed on the mud as it flows through the machine.

The smallest drilled solids can be removed in a *centrifuge* and the fine solids are discarded into the shale pit while the clean mud flows back into the tank (fig. 155).

Figure 155.   A centrifuge removes particles even smaller than silt.

Figure 156.   A mud cleaner is used for mud weighted with barite. It can remove sand without removing the barite from the mud.

Once barite is added to increase the mud weight, the desander, desilter, and centrifuge are no longer used because they will remove barite as well as drilled solids from the mud. For barite-weighted mud, a *mud cleaner* is used to remove most of the solids except barite (fig. 156). Barite is ground to a specific size. Therefore, the solids recovered from a hydrocyclone in the mud cleaner are carefully screened to save any particles that are the same size as the barite. All other-sized particles are thrown away in the shale pit.

After flowing through the gas- and solids-removal equipment, the mud enters a suction tank where the derrickman adds clay, barite, or other chemicals. The purpose of the additions is to make up for the losses that occur as the hole is deepened and to adjust the mud properties as different formations are encountered. Barite is frequently stored in large portable *bulk tanks*. Clay and chemicals are often stored in bags, as shown in the foreground of figure 157.

Figure 157.   Bulk barite tanks with bagged chemicals in the foreground

The derrickman can mix noncorrosive chemicals through the high-velocity *mud hopper* (fig. 158). Corrosive chemicals are added through the closed-top chemical barrel (fig. 159).

Figure 158.   A derrickman, wearing personal protective equipment, adds dry components to the mud through a hopper.

Figure 159.   A closed-top chemical barrel for adding caustic chemicals to the mud in the tanks

Normal drilling operations include drilling the hole and adding a new joint of pipe as the hole deepens. It also involves tripping the drill string out the hole to put on a new bit and then running it back to the bottom (making a *round trip*). Other key steps include running and cementing the large-diameter steel casing used to seal selected intervals of the hole.

DRILLING THE SURFACE HOLE

Engineers create a well plan and a wellbore architecture for every well before it is drilled. A typical wellbore architectural diagram for an onshore well is shown in figure 160. The wellbore diagram shows the hole and casing sizes needed to drill the well to its desired depth.

10
Normal Drilling Operations

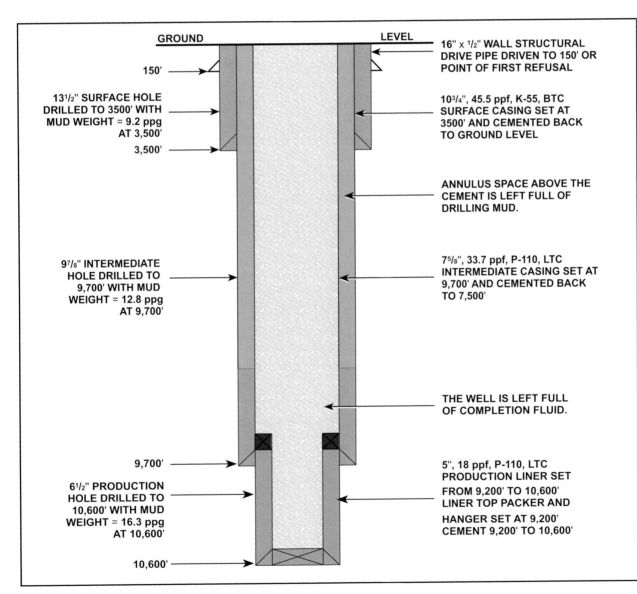

GROUND   LEVEL

150'

13½" SURFACE HOLE DRILLED TO 3500' WITH MUD WEIGHT = 9.2 ppg AT 3,500'

3,500'

9⅞" INTERMEDIATE HOLE DRILLED TO 9,700' WITH MUD WEIGHT = 12.8 ppg AT 9,700'

9,700'

6½" PRODUCTION HOLE DRILLED TO 10,600' WITH MUD WEIGHT = 16.3 ppg AT 10,600'

10,600'

16" x ½" WALL STRUCTURAL DRIVE PIPE DRIVEN TO 150' OR POINT OF FIRST REFUSAL

10¾", 45.5 ppf, K-55, BTC SURFACE CASING SET AT 3500' AND CEMENTED BACK TO GROUND LEVEL

ANNULUS SPACE ABOVE THE CEMENT IS LEFT FULL OF DRILLING MUD.

7⅝", 33.7 ppf, P-110, LTC INTERMEDIATE CASING SET AT 9,700' AND CEMENTED BACK TO 7,500'

THE WELL IS LEFT FULL OF COMPLETION FLUID.

5", 18 ppf, P-110, LTC PRODUCTION LINER SET FROM 9,200' TO 10,600' LINER TOP PACKER AND HANGER SET AT 9,200' CEMENT 9,200' TO 10,600'

COURTESY OF DR. PAUL BOMMER

Figure 160. Typical wellbore architecture

Figure 161. A bit being lowered into the hole on a drill collar

The *surface hole* is a unique part of the well. Blowout preventers cannot be used while drilling the surface hole because the *surface casing* has not yet been placed in the well. The surface casing is the anchor point for all the casing strings and well equipment required to *complete the well*. Although the conductor casing is present, it is not set deep enough to provide a strong anchor point for the blowout preventers and remaining equipment.

The only piece of wellhead equipment sometimes placed on top of the conductor casing is a *flow diverter*. Although a diverter looks like a blowout preventer, all it can do is divert well flow away from the rig. It cannot be *shut in* to stop the flow because the conductor is not set deep enough to withstand shut-in well pressure. Therefore, drilling engineers carefully plan and select the right depth for the surface casing as it is critical for successful well completion.

When the rig-up is complete and the conductor casing is set, the next operation is drilling the surface hole. The large-diameter surface bit drills a hole big enough for the remaining casing strings (shown in figure 160) to be placed in the well. Also, the bit must fit through the conductor casing already in place.

The surface bit is made up on the end of a *bit sub*. The bit sub is a short piece of drill collar with a double box. The box on one end matches the type thread on the pin of the bit, and the box on the other end matches the type thread on the pin of the drill collars. As the hole is drilled deeper, the drill collars, along with the other drilling tools that make up the bottomhole assembly (BHA), are made up on top of the bit sub. The drilling crew uses calipers and a steel measuring tape, or electronic equipment, to carefully measure the diameter and length of each part of the BHA.

The measurements are recorded to determine the depth of the well. The measurements are also important if the BHA should come apart in the well and must be recovered, or fished out, of the well. The record of diameters and lengths of the BHA and drill pipe used on top of the BHA is called a *pipe tally*. The pipe tally book, or automated pipe tally record, is kept in the doghouse at all times.

The bit and the first drill collar are lowered into the conductor hole (fig. 161). There are enough collars and drill pipe made up on top of the first drill collar to lower the bit to the bottom.

On a rig that uses a rotary table and kelly, the driller picks up the kelly and swivel out of the rathole by sliding the open hook on the traveling block through the bail of the swivel (fig. 162). See figure 127 to view the hook and traveling block.

Then, crewmembers stab and make up the kelly onto the top joint of drill pipe that sticks up out of the rotary table. The slips suspend this joint and the entire drill string in the rotary table (fig. 163). With the kelly made up, the driller starts the mud pump and lowers the kelly drive bushing to engage the master bushing.

Figure 162.   A kelly with related equipment in the rathole. The kelly is not visible because it is inside the rathole that extends below the rig floor.

Figure 163.   Red-painted slips with three handgrips suspend the drill string in the hole.

Figure 164.   The kelly drive bushing
is about to engage the master bushing on
the rotary table. The bit and drill string
turn when the driller actuates the rotary
table.

The driller actuates or starts the rotary table to rotate the drill
stem and bit (fig. 164). The driller gradually releases the drawworks
brake, and the rotating bit touches bottom and begins making hole.

On rigs with a top drive, the swivel is part of the top drive, and
there is no kelly. The crew stabs and makes up the last joint of drill
pipe onto the drive stem of the top drive (fig. 165). Then, the driller
starts the mud pump to circulate mud and starts the motor in the
top drive. The string and bit begin to rotate, and the bit is lowered
to the bottom of the hole.

Figure 165.   The motor in the top drive turns the drill stem and the bit.

The weight of the drill string and traveling blocks suspended from the derrick is recorded using an instrument called the *weight indicator*. The black needle, on the inner ring of the indicator, shows the weight suspended from the derrick. In figure 166, the black needle shows the weight at 350,000 lbs (158,800 kilograms). The driller determines the weight placed on the bit by the drill collars by subtracting the entire weight of the drill string before it touches the bottom (the *off-bottom weight*) from the weight shown on the indicator after the bit touches bottom (the *on-bottom weight*). The outer ring with the red pointer can be set to show the weight on the bit, but it must be calibrated first by adjusting the outer ring.

On rigs with a rotary table and kelly, when the bit has made enough hole and the top of the kelly is near the kelly drive bushing, crewmembers say the "kelly is drilled down" (fig. 167). At this point a new joint of drill pipe must be added to the drill string before the hole can be drilled deeper.

On rigs with a top drive, an entire stand of drill pipe is drilled down before a new stand of drill pipe is added. With the kelly (or the stand of drill pipe on top-drive rigs) drilled down, the driller stops rotating, hoists the drill string above the last tool joint, and stops the mud pump. The floorhands *make a connection* by adding (connecting) a new joint (or a stand for top drives) of drill pipe to the drill string so the hole can be drilled deeper.

Figure 166.   The black inner needle on the weight indicator shows the weight suspended from the derrick in thousands of pounds. The red needle indicates the weight on the bit, also in thousands of pounds.

Figure 167.   The kelly is drilled down (close to the kelly drive bushing), and it is time to make a new connection.

To make a connection on a rig with a rotary table and kelly, the driller picks up the drill string high enough for the kelly to clear the rotary table and the last tool joint to be exposed above the floor (fig. 168).

DRILL
PIPE

Figure 168. Using the traveling block, the driller raises the kelly, exposing the first joint of drill pipe in the opening of the rotary table.

Floorhands set the slips around the joint of drill pipe, and the driller slacks off the drill string weight to suspend the drill string in the hole. They latch two big wrenches called tongs on the kelly joint and tool joint of the joint of drill pipe (fig. 169). The tongs act as mechanical hands similar to the way a person screws or unscrews a bolt and a nut. A wire-rope *pull line* runs from the end of the tongs to the breakout cathead on the drawworks.

In figure 169, *breakout tongs* are shown on top around the end of the kelly. The driller engages the automatic cathead and it pulls with enough force to break the connection and begin unscrewing the kelly from the top drill pipe tool joint. To prevent the drill string from turning below the kelly when the breakout tongs are being pulled, a second set of tongs is placed around the tool joint of the top joint of drill pipe. These are called *backup tongs* and are shown below the breakout tongs in figure 169. The backup tongs are secured to one of the derrick legs by a wire rope called a snubbing line. After the tongs are placed on the drill pipe and kelly, but before the driller pulls on the breakout tongs with the automatic cathead, the crew stands away from the arc of the tongs. This is to prevent any crewmember from being struck by a tong in the unlikely event the snubbing line or the pull line breaks.

Figure 169. Crewmembers latch tongs on the kelly and on the drill pipe.

Once the joint is loosened, the driller engages a *kelly spinner*, which is an air motor mounted near the top of the kelly (fig. 170). The kelly spinner rapidly turns, or spins, the kelly to back it out (unscrew it) from the drill pipe tool joint.

Figure 170.   The kelly spinner rapidly rotates the kelly in or out of the drill pipe joint.

Once the kelly is backed out of the tool joint, crewmembers swing the kelly over to the mousehole where the next joint of drill pipe has been placed. They stab the kelly into the joint in the mousehole and the driller spins up the kelly into the joint using the kelly spinner (fig. 171).

Crewmembers grab the tongs, latch them onto the kelly and pipe, and *buck up* (tighten) the joint to the correct torque. To tighten a connection, the tongs are used in reverse of the positions described for breaking out a connection. The breakout tongs will be used as backup when tightening the connection in the mousehole, as seen in figure 171.

Figure 171.   Crewmembers stab the kelly into the joint of pipe in the mousehole. Note the tongs latched onto the joint.

Figure 172.   Crewmembers use tongs to buck up (tighten) one drill pipe joint to another.

Next, the driller uses the drawworks to raise the kelly and the attached joint out of the mousehole. The crew stabs the end of the new joint into the joint suspended by the slips in the rotary table, and, using the kelly spinner and the tongs, they thread the joints together and buck them up to the correct torque. As shown in figure 172, the breakout tongs are on the bottom tool joint to act as a backup, and the *makeup tongs* are on the top tool joint and will be pulled by the makeup cathead on the driller's side of the drawworks.

Finally, the driller picks the drill string weight up off the slips using the drawworks so the crew can remove (or pull) the slips (fig. 173). The driller lowers the newly added joint and kelly until the kelly drive bushing engages the master bushing (fig. 174). The driller starts the pump, begins rotating, and lowers the bit back to bottom to continue making hole (fig. 175).

Figure 173.   Crewmembers remove the slips.   *Figure 174.   The kelly drive bushing is about to engage the master bushing.*

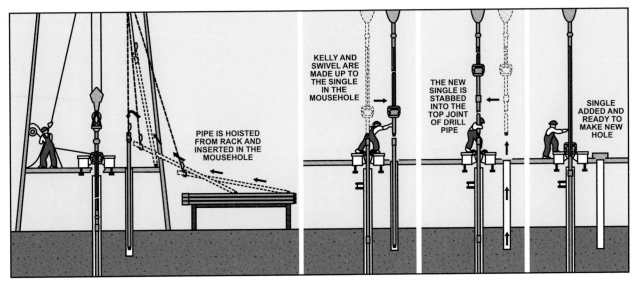

Figure 175.   Making a connection with a kelly

Making a connection on a rig with a top drive is similar to making a connection on a rig with a rotary table. However, with a top drive, the driller has drilled down an entire stand of drill pipe rather than just one joint. The driller raises the drill string off bottom to expose the top tool joint on the drill pipe. Crewmembers set the slips that suspend the drill string in the hole. Although the rotary table is not used to rotate the drill string, it still provides a place for the crew to set the slips (fig. 176).

With the drill string suspended by the slips, the driller reverses the rotation of the top drive to break the connection between the top drive pin and the tool joint box. The backup tongs might be needed to hold the drill string stationary while the top drive is breaking out connections. After the top drive has been unscrewed from the drill pipe, the elevators (which have been tilted out of the way) are latched below the tool joint of the stand in the mousehole (see B in figure 176). Elevators grip the drill pipe below the tool joint and allow the driller to raise and lower the pipe. Then, the driller picks up the stand from the mousehole and allows the elevators to return to vertical.

With the elevators vertical, the crew stabs the new stand into the drill string suspended in the rotary table. The driller lowers the top drive, the elevators slide down the top of the stand, and the top tool joint box moves into a stabbing guide just below the pin of the top drive.

As the top drive continues to be lowered by the driller, the pin end of the top drive is stabbed into the top box of the drill pipe. The driller uses the top drive to spin up the connection of the new stand to the correct makeup torque. Again, it might be necessary to use a

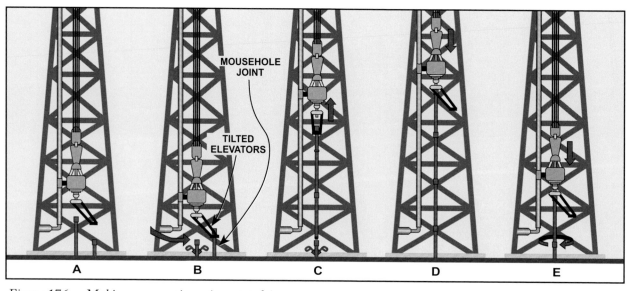

Figure 176.   Making a connection using a top drive

backup tong during the makeup of the new stand. The driller then picks up the string weight off the slips so the crew can pull them from the rotary table, starts the mud pump, begins rotating the string and bit, and lowers the bit to the bottom to continue drilling.

On some rigs, the tongs are replaced with an automatic makeup and breakout machine called an *Iron Roughneck*™ (fig. 177). Once drilling resumes, the crew tallies and places either one new joint or one new stand in the mousehole to be ready for the next connection.

Drilling stops at the depth determined for surface casing in the well plan. At this point, crewmembers trip out (remove) the drill string and bit from the hole so that surface casing can be run in the well and cemented in place. The surface casing provides the anchor point for all remaining well equipment and seals and protects the drilled formations.

Figure 177. An Iron Roughneck™ spins and bucks up joints with built-in equipment.

TRIPPING OUT WITH A KELLY SYSTEM

To trip out on a rig with a kelly system, crewmembers set the slips around the drill stem and break out the kelly. Then they set the kelly, the kelly drive bushing, and the swivel back into the rathole and unhook the blocks from the swivel bail (fig. 178).

Figure 178.   The kelly and swivel with its bail are put into the rathole.

BAIL

SWIVEL

KELLY SPINNER

TOP OF KELLY

RATHOLE

Figure 179.   Crewmembers latch elevators to the drill pipe tool joint suspended in the rotary table.

At this point, the elevators are still attached to the bottom of the hook. The driller lowers the traveling block and elevators down to where crewmembers can latch the elevators onto the pipe (fig. 179). The driller raises the traveling block, thus raising the elevators and pipe, and the floorhands pull the slips. The driller raises the blocks high enough in the derrick to expose one stand of drill pipe. The height of the derrick determines how many joints of drill pipe can be screwed together to form a stand. Although a stand can be one joint, it is generally two or three joints and can be as many as four joints. Pulling the pipe in sections is considerably faster than one joint at a time.

Once a stand has been exposed, the slips are reset around the drill string and the string weight is slacked off onto the slips. The tongs or the Iron Roughneck™ is used to break the connection between the stand and the top tool joint in the slips. A spinner wrench or the rotary table might be used to spin out the last threads of the connection. The driller now picks up the free stand, and the floorhands push the loose end over to a position at the rear of the floor where the stand will be *racked back* (temporarily stored) (fig. 180).

Figure 180.   The floorhands set the lower end of the stand of pipe off to one side of the rig floor.

At the beginning of a trip, the derrickman uses a safety harness and climbs up the mast or derrick ladder to the monkeyboard. The monkeyboard is a small working platform from which the derrickman handles the top of the pipe. Once the floorhands have placed the loose end of a stand on the floor, the driller lowers the block slightly so that the derrickman can unlatch the elevators and pull the top of the pipe back into the *fingerboard* (fig. 181). The fingerboard has several metal projections (fingers) that stick out to form slots into which the derrickman places the top of the pipe. With the top of a stand secured in the fingers of the fingerboard and the bottom of the stand resting on the floor, the blocks are lowered so the elevators can be latched around the top tool joint. This process is repeated until all the drill sting has been pulled from the well.

Figure 181.   The derrickman places the upper end of a stand of drill pipe between the fingers of the fingerboard.

TRIPPING OUT WITH A TOP-DRIVE UNIT

Tripping the drill string out of the hole with a top drive is much like tripping out with a kelly-and-rotary-table system (fig. 182). The main difference is the top drive does not use a kelly and swivel. There is no need to disconnect the kelly and store it in the rathole. Instead, the driller simply unscrews the top drive's driveshaft from the drill string after the floorhands suspend the drill string in the hole with the slips. They use the top drive's built-in elevators to raise the pipe out of the hole and break the stand apart using the same procedure as for a kelly-equipped rig.

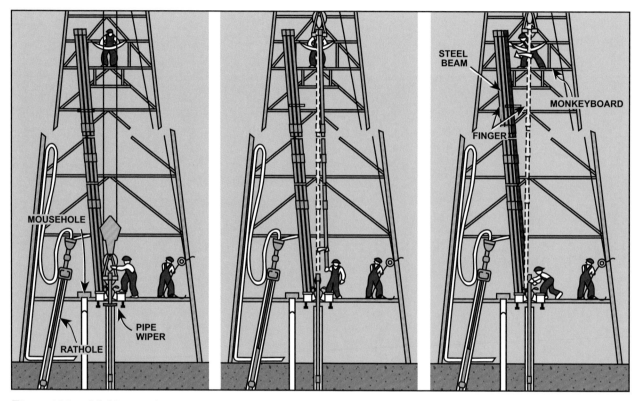

Figure 182.   Making a trip

TRIPPING OUT WITH A PIPE RACKER

Some rigs have an automated device to run pipe in and pull pipe out of the hole. It is an automatic pipe-handling device called an *automatic pipe racker* that moves on a track on the rig floor beneath the fingerboard (fig. 183). The driller usually operates the automatic pipe racker from the rig floor. To pull pipe out, the driller uses the drawworks to raise the pipe, stops it at the desired height, and applies automatic slips to suspend the string.

Next, the driller activates a control to move the pipe racker into position near the tool joint to be broken out. Automatic breakout tongs built into the racker loosen the joint, and a built-in automatic spinner spins the joint apart. The operator retracts the racker's tongs

Figure 183.   Top view of an automatic pipe handling device manipulating a stand of drill pipe

and extends two arms from the racker. One of the arms grips the pipe near the rig floor and the other grips it higher up, just beneath the fingerboard. Gripping the pipe and moving the racker in its track puts the bottom of the stand or joint on the rig floor and places the top of the stand or joint into the fingerboard.

With the stand or joint *set back* in the mast or derrick, the driller retracts the racker's arms and moves the racker into position for the next stand or joint to be pulled from the hole. The driller extends the racker's automatic tongs and breaks out the next stand or joint of pipe. The racker then sets back the stand or joint in the fingerboard. The driller repeats this process until all the pipe is out of the hole.

The automatic pipe racker is primarily a safety device because it does not require the crew to handle the pipe. However, the pipe racker is usually not as fast as the crew in terms of making a trip.

RUNNING SURFACE CASING

In the well plan, the engineers specify the size and properties of the surface casing required for the job. Running casing into the hole is similar to running drill pipe, although the casing diameter can be much larger and requires special elevators, slips, and tongs to fit it. The rig might not have the equipment necessary to run the casing so the operator hires specialists known as a casing crew. The casing crew supplies the correct size and type of elevators, slips, and tongs calibrated to make the casing up to the correct torque (figs. 184 through 189). Some rigs, especially offshore, will have all the equipment required to *run casing* and generally do not require the special casing crew.

Figure 184. A casing crewmember cleans and inspects the casing as it lies on the rack next to the rig.

Figure 185. Casing threads have been cleaned and inspected. A fresh coat of lubricant and thread sealer (pipe dope) will be applied before the casing joints are made up on the floor.

Figure 186.   A joint of casing being lifted onto the rig floor

Figure 187.   A joint of casing suspended in the mast; note the centralizer

Figure 188.   Casing elevators suspend the casing joint as the driller lowers the joint into the casing slips. The slips suspend the casing string in the hole.

STABBING
BOARD

*Figure 189.   Working from a platform called the stabbing board, a
casing crewmember guides the casing elevators near the top of the casing
joint.*

The casing crew installs extra equipment onto the casing to
improve the cementing process that secures the casing in the well.
A *guide shoe* is placed on the bottom of the first joint of casing to be
placed in the well. The guide shoe has a rounded profile that allows
the end of the casing to slip past small ledges in the well or easily
slide around deviated portions of the well (figs. 190 and 191).

A *float collar* is then placed one or two joints above the guide
shoe (figs. 192 and 193). This is a short piece of pipe, the same size
as the casing, that has a *flapper* or spring-loaded dart that prevents
cement from flowing back inside the casing at the end of the cementing job.

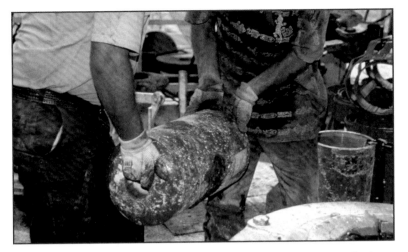

Figure 190.   Crewmembers lift the heavy steel-and-concrete guide shoe.

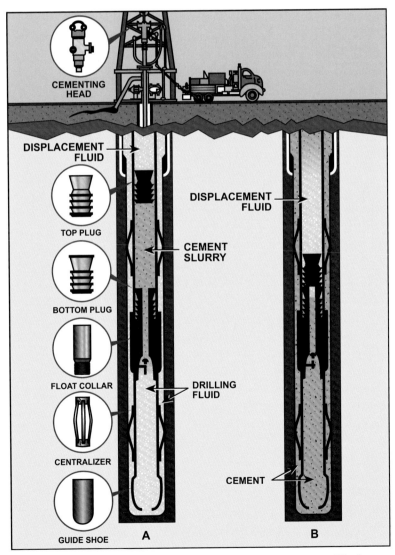

Figure 192.   Cementing the casing: (A) the job in progress; (B) the finished job

Figure 191.   The guide shoe is made up on the bottom of the first joint of casing to go into the hole.

Figure 193.   Crewmembers install a float collar into a casing string.

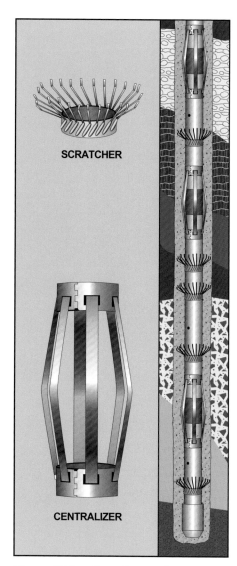

Figure 194. Scratchers and centralizers are installed at various points in the casing string.

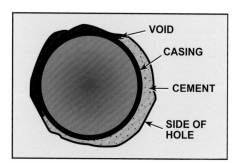

Figure 195. Top view of casing that is not centered in the hole. A mud-filled channel remains where the casing touches the side of the hole. The mud-filled channel will not be sealed.

Both the float collar and the guide shoe are made of drillable material. The next bit used can drill out the inside of this equipment, and the hole can be drilled deeper below the casing setting depth. *Centralizers* will be strapped around the outside of the casing to keep the casing centered in the middle of the borehole (fig. 194). This allows cement to fill up evenly around the outside of the casing and not leave any gaps in the cement (fig. 195). *Scratchers* (shown in figure 194) can also be strapped on the outside of the casing to scratch away some of the mud cake to improve contact between the cement and the hole.

CEMENTING

Cement supports and protects the casing and bonds it to the hole. The cement also seals the annular space between the casing and the hole, preventing fluids in one formation from migrating to another. The operator hires an oilwell cementing company to perform this job. Cementing companies stock the special equipment needed to transport the various types of cement to the well. At the well, the company mixes the dry cement with water to form a *slurry* that can be easily pumped. The cement mixers continuously blend the water and cement to make a uniform mixture as the cement pumps move it down the casing and into the *annulus*.

High-pressure pumps move the slurry through steel pipes or lines to a *cementing head*, or plug container (see figure 192). The cementing head is mounted on the topmost joint of casing hanging in the mast or derrick (fig. 196).

Just before the slurry arrives at the head, a crewmember releases a rubber plug, called a *bottom plug*, from the cementing head. As seen in figure 192, the bottom plug separates the cement slurry from any drilling fluid inside the casing and prevents the mud from contaminating the cement. The slurry moves the bottom plug down the casing. The plug stops, or *seats*, in the float collar. Continued pumping breaks a membrane on the bottom plug and opens a passage. Slurry then goes through the bottom plug and continues down the last few joints of casing. It flows through an opening in the guide shoe and up the annular space between the casing and hole. Pumping continues until the slurry fills the annular space.

As the last of the cement slurry enters the casing, a crewmember releases a second plug, called a *top plug*, from the cementing head (refer to figure 192). A top plug is similar to a bottom plug except it has no membrane or passage. The top plug separates the last of the cement to go into the casing from *displacement fluid*. Displacement fluid (drilling mud or water) moves the cement from the casing, as the cement pump applies pressure to move the cement, now separated from the displacement fluid by the top plug, down the casing.

The top plug seats on, or *bumps*, the bottom plug in the float collar. When it bumps, there is a noticeable increase in pumping pressure because the plug is solid. At this time, the pump operator shuts down the pumps. Cement is only in the casing below the float collar and in the annular space. Most of the casing is full of displacement fluid (refer to figure 192). It is critical that the surface casing have the cement fill the annular space from the bottom of the casing to the top of the ground or, if offshore, to the seafloor. The surface casing is the *anchor point* for all the remaining equipment to be used in the well, and it must be securely sealed and supported in the ground or seafloor.

After the cement company pumps the cement and removes its equipment, the operator and drilling contractor must wait for the cement to harden. This period of time is referred to as *waiting on cement*, or *WOC*. Usual WOC time is a few hours, depending on the cement formulation and the well temperature.

After the cement hardens, the first section of the wellhead is installed and the BOPs are placed or *nippled up* on top of the wellhead. The BOPs, the wellhead, and the inside of the casing are pressure-tested to make sure there are no leaks. After this, a smaller bit is made up on the BHA as detailed in the well plan. The new bit is lowered into the casing and used to drill out the float collar, the cement remaining in the float joints of casing, and the guide shoe. The cement seal at the bottom of the casing on the outside is pressure-tested, and drilling the new hole resumes.

The same procedure is used for land rigs and bottom-supported rigs. However, floating rigs must use a different procedure to achieve the same goals. The primary difference on floating rigs is that the wellhead equipment including the BOPs are placed on the seafloor and connected back to the vessel with a marine riser.

Figure 196.   A cementing head (plug container) rests on the rig floor, ready to be made up on the last joint of casing to go into the hole.

TRIPPING IN

The process of returning a bit to the bottom of the well is called tripping in. To trip in, the derrickman returns to his position on the monkeyboard. The driller uses the elevators to raise the blocks to the height of the derrickman. The derrickman takes the top of a stand out of the fingerboard and latches it in the elevators. The driller raises the blocks enough to pick the stand up off the floor. The floorhands steady the stand and position it over the hole. The bit and first stand of the BHA are lowered into the well and secured by the slips. The driller returns the blocks and elevators to the position of the derrickman, who latches another stand in the elevators. The floorhands stab this stand of pipe into the suspended stand (fig. 197).

The crewmembers spin up the stand with a spinning wrench (fig. 198). With the pipe spun up, the crew uses the tongs to buck up the joint to the correct torque (fig. 199). This process is repeated until the bit is returned to the bottom of the well.

Figure 197.   To trip in, crewmembers stab a stand of drill pipe into another.

Figure 198.   After stabbing the joint, crewmembers use a spinning wrench to thread the joints together.

Figure 199.   After spin up, crewmembers use the tongs to buck up the tool joints to the correct torque.

DRILLING AHEAD

Drilling proceeds to the next depth as determined in the drilling plan. More than one bit might be required to drill a certain section of the well. If a bit becomes dull and must be replaced, another round trip is required.

The next step in the drilling plan might be the end of the well where oil- and gas-bearing rocks are believed to exist, or it might be an intermediate depth. At an intermediate depth, another *casing string* is required to protect the formations drilled from higher or, in some cases, lower pressures that are anticipated at deeper depths. If *intermediate casing* is required, it is run and cemented in a similar way to the surface casing. The intermediate casing might be run all the way back to the surface or suspended from a hanger at the bottom of the last casing string as a liner.

When the hole reaches the *total depth (TD)*, the well is evaluated to determine if sufficient oil and gas exist in the rock to warrant completion. If a completion is warranted, *production casing* is run and cemented in the well to stabilize and seal the bottom portion of the well.

Formation evaluation is the process used by operators to determine if rock layers contain hydrocarbons. Formation evaluation can determine if sufficient quantities of hydrocarbons are present and if the rock has enough permeability to allow a commercial completion. The techniques addressed in this chapter are the examination of cuttings and drilling mud, well logging, drill stem testing, and coring.

11
Formation Evaluation

EXAMINING CUTTINGS AND DRILLING MUD

One of the oldest formation evaluation techniques is to simply look at the cuttings and the drilling mud returning from the bottom of the hole (fig. 200). A geologist or trained technician who examines the returning drilling mud and cuttings is called a *mud logger*.

The rock type can be identified from the cuttings. This is important because reservoirs typically fall into broad categories by rock type. For example, reservoir rocks are often sandstone and limestone, which develop the correct combination of porosity and permeability needed to contain hydrocarbons and allow them to flow. A rough idea of the porosity of a rock can be determined by viewing cuttings under a microscope. If a rock contains oil, trace amounts of oil will coat the cuttings even after they have been circulated in drilling fluid and brought to the surface.

Oil is a polarizing compound. It will have a fluorescent shine when viewed in a black light box. The oil stain on cuttings can be confirmed by flushing the oil off the cuttings with a solvent. The streaming solvent will also fluoresce under the black light. In this way, an oil stain can be differentiated from other rock mineral that might also fluoresce. Using this method to determine the presence of oil does not work if an oil-based mud is used as a drilling fluid.

Figure 200.   A handful of cuttings made by the bit

If gas is in the rock, some gas will be released into the mud when the rock is being drilled. The gas will be carried to the surface in the drilling fluid and can be detected by combusting the gas as it flows over a heated filament or *hot wire*. The filament is attached to an electric circuit that changes resistance in proportion to the amount of heat given off. The device is calibrated against a methane-and-air mixture, and the electrical output is scaled in gas units. It is often possible to obtain *chromatographic analysis* of the gas to determine its various components. There are almost always very low levels of gas in mud. This gas appears in the analysis as only a few gas units. The low level of gas becomes a background-level gas against which any increases can be easily seen.

The drilling rate can be recorded by the mud logger. Porous and permeable rocks often drill faster than less porous and permeable rocks. If drilling rate increases, it might be because the bit is drilling a rock with these improved characteristics. Rocks with higher porosity and permeability are good reservoir rocks that might contain hydrocarbons.

The well depth when the cuttings and gas samples reach the surface is not the depth at which they were drilled. The well has gotten deeper as the cuttings circulated to the surface. To record the cuttings and gas units at their actual depth, the mud logger will occasionally place some carbide in the drill pipe during a connection. The carbide reacts with water in the mud and gives off an acetylene gas that can be identified by the gas detector when it returns to the surface. The time needed for the acetylene gas to return to the surface can be used to calculate the time delay, or *lag*, for the cuttings and gas samples to reach the surface from the bottom of the well. The lag is used to compute the correct depth when the cuttings and gas samples were drilled, as opposed to the well depth when they arrived at the surface.

The drilling rate, rock type, gas units, gas analysis, and sample description are recorded on a chart known as a mud log. A section of a mud log is shown in figure 201. Mud logs are an accurate indicator of the presence of hydrocarbons and a reliable guide regarding the quality of the rock holding the hydrocarbons. Although other techniques are now also used to supplement the mud log, historically, more oil and gas have been found by observing the cuttings and mud than by any other way.

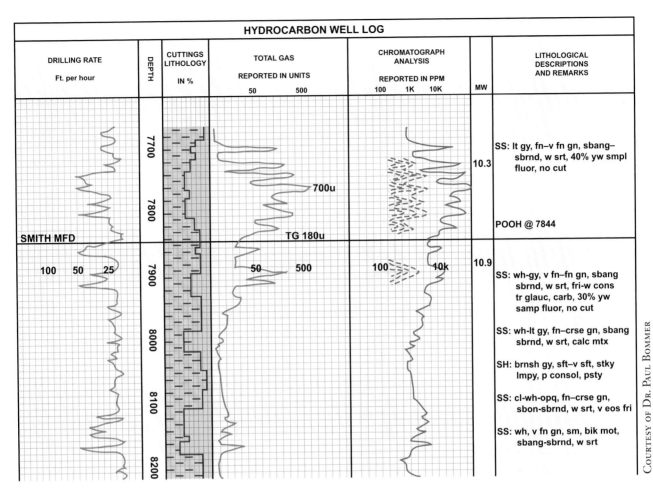

Figure 201.   Mud log section showing a formation that contains hydrocarbons

WELL LOGGING

Well logging is a widely used evaluation technique, and many kinds of logging tools are available. Some tools measure and record the natural and induced nuclear, or radioactive, attributes of a rock. Logs of this type include the *gamma ray log*, which indicates how much of a layer is made up of naturally radioactive shale. The *neutron log* and *density log* measure radioactive response, indicating the porosity of a layer. *Electric logs* measure the way in which formations respond to an electric current. The electric resistivity of a layer is a combined indicator of the type of rock present and the kind of fluid it contains. A *sonic log* indicates the porosity of a layer by measuring the speed with which sound travels through a formation. Running several of these well logs provides engineers and geologists with effective information to determine the rock types that have been drilled and the fluids contained in the rocks.

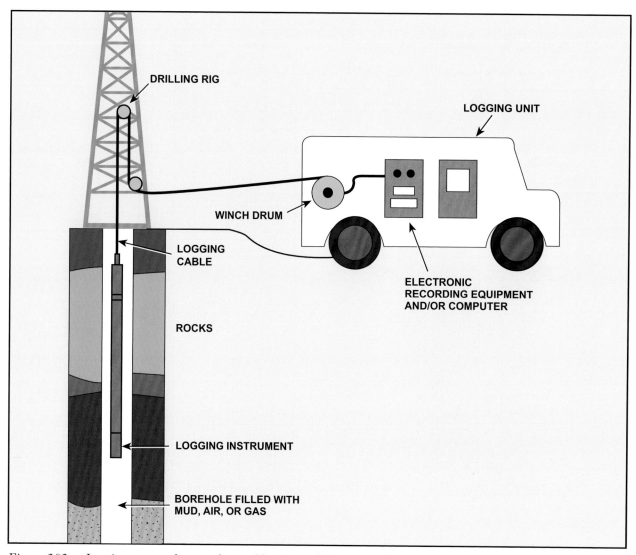

Figure 202.   *Logging personnel run and control logging tools by means of wireline from a logging unit.*

Well logs are created and measured by a logging service company. The well logging company provides a truck-mounted logging unit for land rigs or a skid-mounted unit on offshore rigs. From the well logging unit, the logging tools are lowered into a well on an electric *wireline* (fig. 202).

The tools are lowered to the bottom of the hole and then slowly reeled back up using a hoist. When activated, the tools measure the rock properties and transmit the data to a logging unit where it is recorded by a computer. The logging data is available for display on the computer monitor or to be printed out (fig. 203). The data can also be transmitted electronically to other locations. By carefully examining the well logs, engineers and geologists can determine the rock types and the fluids in the rocks. Based on this information, they will decide whether or not to complete the well or whether to obtain more data.

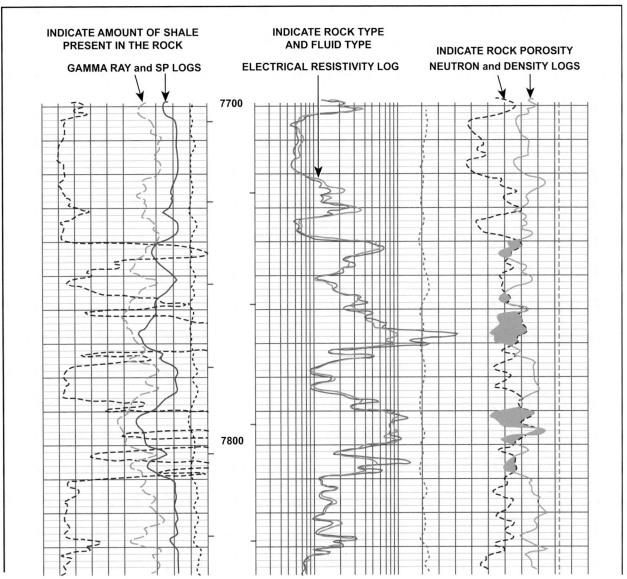

INDICATE AMOUNT OF SHALE
PRESENT IN THE ROCK

GAMMA RAY and SP LOGS

INDICATE ROCK TYPE
AND FLUID TYPE

ELECTRICAL RESISTIVITY LOG

INDICATE ROCK POROSITY
NEUTRON and DENSITY LOGS

7700

7800

COURTESY OF DR. PAUL BOMMER

Figure 203.   A well-site log is interpreted to give information about the formations drilled.

During drilling, the operator can run *logging while drilling (LWD)* tools in the drill string. These instruments incorporate sophisticated electronic devices that sense, transmit, and record formation rock data as the bit drills ahead. The LWD tool transmits information about a formation using a tool that creates a pulse in the drilling mud. The pulses transmitted through the mud send formation information to computers on the surface. The computers convert the transmitted code into data and display the well logs for engineers and geologists to study.

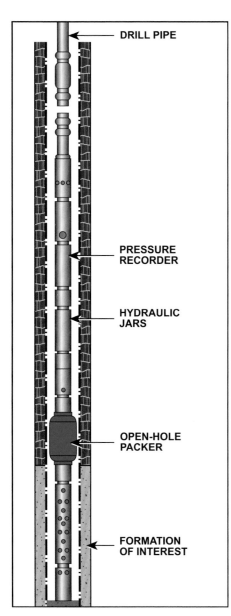

Figure 204. Drill stem test tools

DRILL STEM TESTING

To further determine the potential of a producing formation, the operator might order a *drill stem test*, or *DST*. The DST crew makes up the test tool on the bottom of the drill stem and then lowers the tool to the bottom of the hole (fig. 204).

The crew applies weight to the drill string test tool to expand a hard-rubber sealing element called a *packer*. The hole above the expanded packer is sealed off. Then, the DST crew opens ports in the tool that lies below the packer. The formations below the packer are free to flow into the well once the ports are open. If a hydrocarbon reservoir is present, the oil, gas, and water flow rates and the flowing pressures can be measured. The fluids can be sampled and accurate flow rates measured, if the reservoir can flow fluids to the surface. If the reservoir does not flow all the way to the surface but allows some hydrocarbons to flow into the drill pipe, the reservoir fluids become trapped inside the closed tool and samples can be obtained. Drill stem testing can provide the most graphic demonstration of the production capability of a reservoir (fig. 205).

A wireline device called a *repeat formation tester (RFT)* can be used to obtain a small volume of reservoir fluid and measure reservoir pressures (fig. 206). The well logging company provides the RTF equipment, which is run on an electric wireline. Once at the correct depth, the tool is forced over to the side of the hole by an electrically operated piston. When the tool face is against the wall of the hole, a small snorkel is extended and the tool is opened. A reservoir fluid sample will flow into a container when it is opened. The reservoir pressures are recorded simultaneously with the flow period. When the tool is unset and retrieved, a reservoir fluid sample is obtained.

The main drawback to DSTs and RFTs is that the tools must remain stationary and in contact with the sides of the borehole. In soft formations or formations with low reservoir pressure, it is possible the tools might become stuck. If this occurs, a *fishing* job must be attempted to recover the tools before the hole can be used. As discussed in the section on Fishing in Chapter 13, Special Operations, these operations are time consuming, expensive, and not always successful.

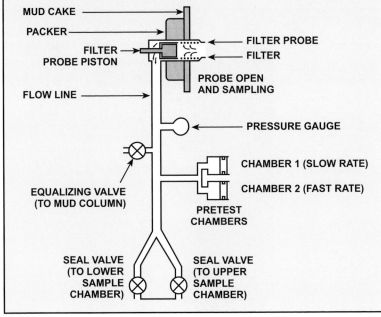

Figure 205.   *A successful DST*

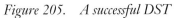

Figure 206.   *Repeat formation tester (RFT) tool*

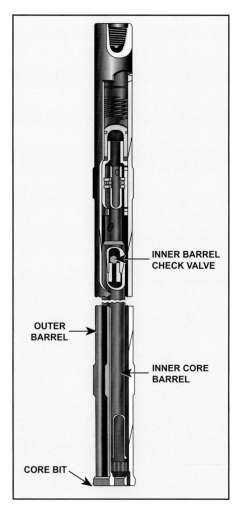

Figure 207a.   A whole core barrel

CORING

A *core* sample can be taken if the operator wants to directly examine a formation sample larger than small cuttings. Two coring methods are *whole coring* and *sidewall coring* (fig. 207a and 207b).

To take a *whole core*, the drilling crew makes up a *core barrel* and bit and runs it to the bottom on the drill string. When the driller rotates the core bit, it cuts a cylinder, or core, of rock. The core is often several inches (or millimetres) in diameter and several feet (or metres) long. As the core barrel cuts the core, the core moves into a tube in the barrel.

After the desired length of core is cut or the barrel is full, the crew trips out the drill string and core barrel. At the surface, the core is removed and shipped to a laboratory for analysis. Figure 208A shows a segment of oil-saturated core taken from a well in South Texas.

Sidewall cores are taken using a sidewall core gun, which is a series of hollow metal tubes, each loaded in front of an explosive charge. The gun is lowered to the desired depth on an electric wireline. The logging engineer fires one core barrel at a time. The barrel is driven into the side of the formation but is still connected to the gun by steel cables. When the gun is moved, the cables pull the tube containing a rock sample out of the formation. As many as twenty-five samples can be taken with each gun run into a well, and it is possible to run several guns in tandem. Individual samples are much smaller than the whole core. A sidewall core is only about one inch (25.4 mm) in diameter by two or three inches (50.8 mm to 76.2 mm) long (fig. 208B). The sidewall cores are pressed out of the barrels, placed in plastic containers labeled with the depth at which the sample was taken, and shipped to a core laboratory for analysis.

Figure 207b.   Sidewall coring device

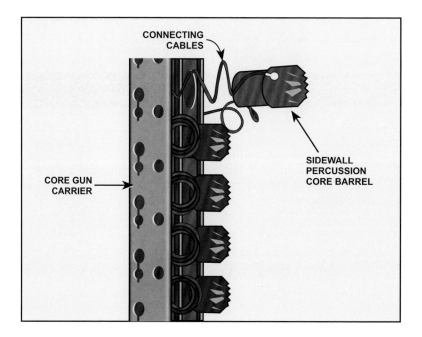

Both types of cores have a drawback because they have been contaminated by drilling fluid. The reservoir fluids in the cores do not indicate the true saturations existing in the reservoir, because they have been partially flushed by drilling fluid during the coring and drilling process. The whole core is a more representative sample of the reservoir than a sidewall core because it is longer and larger in diameter. A whole core might be long enough to actually see the entire vertical thickness of the reservoir. The depth at which the whole core is taken must be accurately identified because the core bit is drilling this part of the well while the core is being taken. It is possible that part of the reservoir might be missed by the core if the depth to the top of the reservoir has been chosen incorrectly.

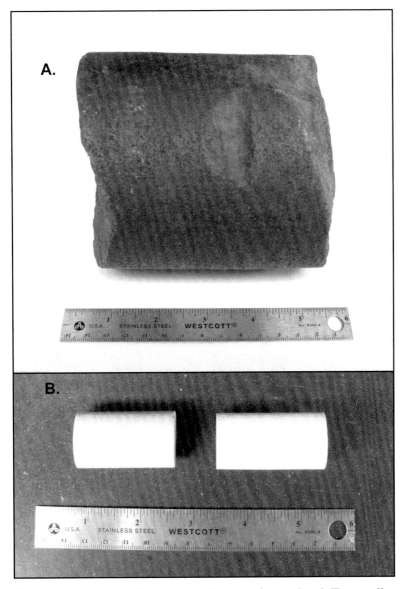

Figure 208. A. An oil-saturated whole core from a South Texas well; B. Sidewall cores

It is more time consuming to obtain whole cores rather than sidewall cores because two round trips are involved in running and retrieving the core barrel. The sidewall cores take less time because running wireline is faster. They can be placed with great accuracy after the well logs have provided a picture of the reservoir. The rock in a sidewall core has been somewhat altered by the act of driving the barrel into the rock. The rock might be compressed or fractured during the process. Sidewall cores are useful for relative comparisons. They are not nearly as representative as a whole core. To obtain a detailed reservoir analysis, the whole core sample is preferred.

12

Completing the Well

Once the formation evaluation is done, the operator must decide if the well should be completed as a producing oil or gas well. If the well does not contain hydrocarbons, or not enough to pay for the completion, the well will be *plugged and abandoned (P&A)*.

PLUGGING AND ABANDONING A WELL

To P&A a well, the drilling rig pumps several cement plugs through the drill pipe. The cement plugs are used to isolate and seal unprofitable hydrocarbon zones from nonhydrocarbon-bearing zones and to seal freshwater zones from saltwater-bearing zones. The intervals between cement plugs are left full of drilling mud. At this time, it might be possible to cut off and recover some of the intermediate casing string (if one is present) for use in other wells. The surface casing string is always left in place and sealed at the bottom and top by either cement plugs or a combination of mechanical and cement plugs. The surface casing will be cut off below the ground level or mud line and a *cap* placed on the stub. If the well is on land, the well site will be environmentally restored after the drilling rig has been moved off the location.

COMPLETING A PRODUCING WELL

The drilling rig is used to run and cement production casing as described previously. The blowout preventers are removed and a production wellhead is attached to the top of the casing. The production wellhead seals the tops of the various casing strings in the well, provides a place to suspend and seal p*roduction tubing* as needed, and provides the valves that control flow out of the well. Figure 209 shows a typical land wellhead or *Christmas tree*.

Figure 209. This collection of valves and fittings is a Christmas tree.

Figure 210.   Subsea wellheads

Examples of subsea wellheads are shown in figure 210. Offshore, the drilling rig will be used for completion operations. On land, the drilling rig is removed at this point and a smaller, less-expensive rig known as a *completion rig* is used to finish the completion. In special cases, no rig is required to complete the well.

PRODUCTION TUBING

Production tubing can be used in completing a well. The tubing is a small-diameter string of jointed pipe that fits inside the production casing. The bottom of the tubing can be sealed inside the production casing using a packer. The tubing provides a flow conduit for reservoir fluids to the control valves on the Christmas tree. It also protects the production casing from corrosive fluids and high pressures and temperatures that could damage the production casing. The smaller-diameter tubing might allow the reservoir to flow more efficiently, or it can be used to install *artificial lift* hardware if the well stops flowing naturally.

In all offshore wells, the tubing has a *subsurface safety valve (SSSV)* that automatically closes in the event of a severe storm or other disaster. The SSSV prevents the uncontrolled flow of hydrocarbons from the well if the platform, production facility, or wellhead is damaged or lost. Some onshore wells might not require production tubing if the reservoir fluids are noncorrosive and can flow up the larger diameter production casing.

It is possible to run a continuous string of tubing into a well. This type of tubing is called *coiled tubing* because it is not jointed and is delivered to the site coiled on a large reel (fig. 211). The tubing can be installed without a rig and can be hung in the wellhead for permanent use, or it can be reeled back out of the well on the large spool. Coiled tubing is frequently used to clean debris from inside a well or to convey treating fluids to the bottom, then it is reeled back out of the well.

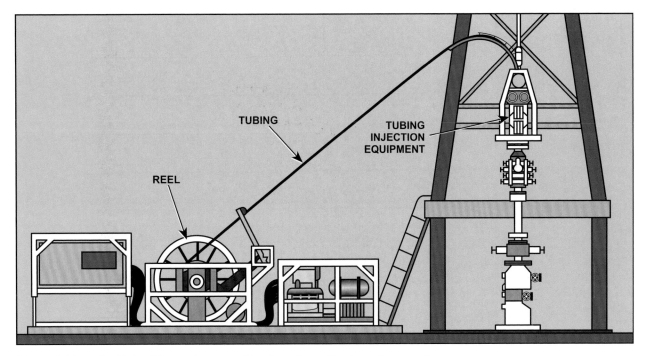

Figure 211. A coiled-tubing unit runs tubing into the well using a large reel.

PERFORATING

The producing formation might be left open by setting the production casing at the top of the reservoir. This style is called an *open-hole completion*. For this type of completion to be effective, the reservoir must not collapse during production. Most reservoirs have the production casing or *liner* run across the productive section and cemented in place. The production casing will support the formation and prevent collapse of the well, and, if cemented, will seal and isolate the reservoir.

Once the production casing is in place, the reservoir fluids cannot flow into the well until holes are made in the casing to reconnect the reservoir with the inside of the well. The holes are called *perforations* (fig. 212).

Figure 212.   Perforations (holes)

Perforations are made by detonating a *shaped charge* at the correct depth. The shaped charges are placed in a carrier called a *perforating gun* (fig. 213). When fired, the shaped charges form a high-velocity jet of energy, creating a hole through the casing, the outside cement, and a tunnel some distance into the formation.

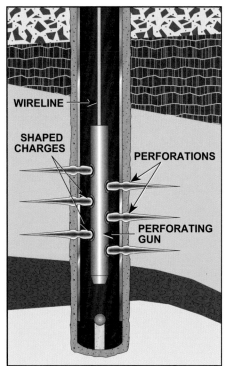

Figure 213.   Shaped charges in a perforating gun make perforations.

The gun can be run into the well on an electric wireline or transported on the bottom of the tubing. The shaped charges can be placed in various shots-per-foot in a selected pattern spiraling around the gun. The gun is positioned at the correct depth in the well, typically a gamma ray log is used because it is unaffected by the presence of the casing. The charges are fired using a blasting cap and primer cord. After the gun has fired, it is retrieved from the well by reeling in the electric line. If the gun was conveyed on the tubing, it can be retrieved either by pulling the tubing out of the well or releasing it from the tubing and allowing it to fall to the unused portion of the well at the bottom, known as the rathole.

WELL TESTING AND TREATING

After perforation, the well is flowed to determine the rate and pressure at which the oil and gas can be produced to the surface. In some reservoirs, the flow rates and pressures can be enhanced by *treating* or *stimulating* the reservoir using various methods.

Acidizing

An acid mixture can be used to dissolve part of the reservoir, thereby increasing the permeability of the reservoir near the wellbore. Limestone is rapidly dissolved in *hydrochloric acid (HCL)*. Some clay and sandstone minerals can be dissolved in *hydrofluoric acid (HFL)*, although such reactions are much slower than the limestone-HCL reaction. If the permeability of the reservoir is increased, the flow rate will also increase. The acid volumes used and the reaction rate of the acid with the rock will generally affect only that part of the reservoir near the wellbore. Acid, used in this way, is only useful in removing low permeability or damaged areas very close to the well.

Fracturing

Hydraulic *fracturing* is used for reservoirs with extremely low permeability or reservoirs that are too deeply damaged for acid to repair. Fluids are pumped into the well at pressure sufficiently high to open the reservoir cracks. More fluid pumped into the well causes the crack or fracture to grow longer, wider, and perhaps higher. After the fracture is created, it is filled with *proppant* used to prop the fracture open and keep it from closing after pump pressure is removed. The proppant can be simple sand, bauxite, or man-made ceramic beads, depending on the depth and properties of the reservoir. Pathways in limestone can be etched using acid as the fracturing fluid. The fracture becomes a high-permeability channel from the reservoir to the wellbore. Correctly designed, fracturing jobs can greatly increase the natural flow rate of a reservoir.

Gravel Packing

Some sand reservoirs are unconsolidated. In reservoirs of this type, the sand grains are either poorly cemented to each other or, in extreme cases, not cemented at all. In these cases some reservoir sand also flows into the well when the flow rate of reservoir fluids becomes sufficiently high. Sand can plug the inside of the well. It can also abrade the tubing, casing, and wellhead and leave a void outside the casing that might cause the well to collapse. To prevent the reservoir sand from flowing into the well, a *gravel pack* is installed. A gravel pack puts carefully sized sand (gravel) between the reservoir and the outside of a screen placed in the well. The size of the gravel prevents all but the finest of reservoir sand from flowing through the gravel pack. The screen has slots small enough to prevent the gravel from flowing back into the well (fig. 214). The fine reservoir particles that might flow through the gravel and screen with the reservoir fluids are small enough in size and volume to not harm the *well completion*.

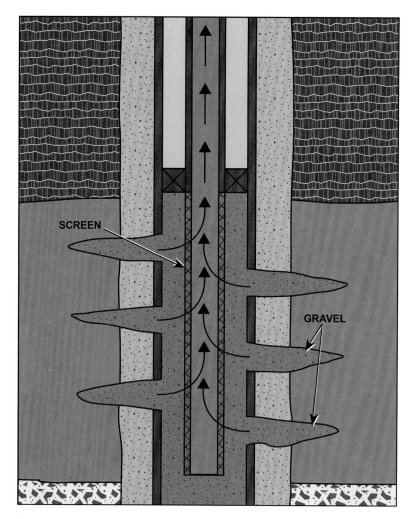

Figure 214.   A gravel pack

There are several special operations used in oilwell drilling: directional drilling, fishing, and *well control*.

Special Operations

DIRECTIONAL DRILLING

No well is ever perfectly vertical. Even wells meant to be drilled vertically will wander a few degrees from vertical and move in different directions. Routine measurements are taken during drilling to determine if a well is deviating from vertical by more than the allowed amount (normally less than 5 degrees). If so, careful drilling practices, such as changing the placement of stabilizers in the BHA or adjusting the *rotary speed* or weight on bit, will bring the well back within the tolerances normally allowed for vertical wells.

Directional drilling is used when a well is intentionally deviated to reach a *bottomhole location (BHL)* that is different from the *surface location (SL)*. Directional drilling is done for many reasons. The BHL might be under an obstruction such as a building or lake where rigging up over the required BHL is not possible. It might be necessary to drill several wells from a fixed place, such as an offshore platform or an onshore drilling island (fig. 215), to different bottomhole locations.

Part of an existing well might become blocked with lost drilling tools that are unrecoverable, or a well might have been drilled into an unproductive part of the reservoir. It is possible to set a plug in the lower part of the well and deviate, or *kick off*, the well to a new BHL. Some reservoirs are more efficiently produced by wells drilled at a very high angle. These wells are known as *horizontal wells* because the inclination angle from vertical reaches 90 degrees or more.

Older directional drilling methods placed inclined wedges, called *whipstocks*, in the well to force the bit to move in the desired direction. In soft sediments, it is possible to place a large bit nozzle or jet in the desired direction and simply erode the well's starting path. Although time consuming, these methods are still used at times.

The two faster and often more reliable methods of directional drilling are:

- *Slide drilling* with a motor
- Drilling with a *rotary steerable assembly*

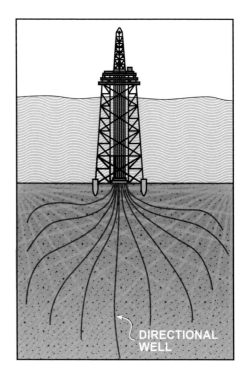

Figure 215. Several directional wells tap an offshore reservoir.

Slide Drilling with a Motor

Current downhole motors can have a flexible section on the end where the bit is attached. The flexible part of the motor is called a *bent housing*, and it allows a small angle to be fixed on top of the bit. The small angle, or bend, allows the bit to be pointed in the desired direction. Once pointed, the drill string is not rotated. The bit is made to rotate by the mud flowing through the motor. When the bit drills new hole, the drill string slides along behind in a style known as slide drilling. Slide drilling is a slow process compared to drilling with the rotating drill string because there is substantial friction between the drill string and the hole while sliding.

A guidance tool can measure the direction of the hole using accelerometers and magnetometers. The guidance tool is powered by batteries and sends signals through a transmitter data back to the surface. The tool is placed in a nonmagnetic drill collar so that true readings can be obtained. The data signals are sent as pressure pulses back up the mud-filled annulus. At the surface, a pressure transducer records the data pulses and sends them to a computer where they are decoded and display the direction of the bottom of the hole. This system is called *measurement while drilling (MWD)*. The MWD tool will not work if the well is being drilled with a gas or foam or, if the annulus is not continuously full of liquid. For the special cases where MWD cannot be used, other tools called steering tools can be used in combination with an electric wireline to determine the direction of the well.

Alternate methods of MWD include electro-magnetic (EM) signals and electric signals. The EM system sends radio waves through the earth that can be detected by antennas placed around the rig. The EM system does not work well in some parts of the world and offshore. Using the electric system, an electric signal can be sent directly from the tool to the surface if intelligent drill pipe is used. An intelligent drill pipe has an electric wire imbedded in the steel that provides a direct link to the tool. The only problem with the electric system is that it requires specially-made drill pipe. The downhole tools that send and receive signals from the electric wire or the EM system must be custom made to be compatible, just as is the older mud pulse tool.

It is not necessary to continue slide drilling after the path of the hole is changed to the desired direction. The drill string can be rotated to increase the drilling rate and new direction measurements are made to ensure the hole is going in the right direction. A new slide is performed to move the hole back in the desired direction if the hole does not maintain the desired path.

Rotary Steerable Assemblies

Two drawbacks of slide steering are: it is time consuming to orient the bit, and slide drilling is not as fast as drilling with a rotating drill string. A rotary steerable assembly used to remedy slide drilling drawbacks is a recent technology still in development.

A rotary steerable assembly has a computer programmed with the desired wellbore path and one form of several mechanisms that allow the bit to be pointed or pushed in the desired direction while the drill string is rotated. Directional measurements are recorded in the memory of the assembly for verification and adjustment of the hole path. These tools allow the well to be drilled faster because the drill string is rotated, reducing the friction associated with slide drilling. They also eliminate the frequency and time associated with taking MWD directional readings and reorienting the bit. The drawbacks to the rotary steerable assemblies are that they are more expensive than the slide drilling tools and, if the directional system malfunctions, the hole might be drilled hundreds of feet (metres) before the error is noticed.

FISHING

A piece of equipment, a tool, or part of the drill string lost in the hole is commonly called a *fish*. Drilling personnel also call small pieces, such as a bit cone or a tong die, *junk*. When a fish has been left in the hole, the crew must remove it, or fish it out, because the drill bit cannot get past the fish to continue drilling. Great care is taken to avoid leaving junk or a fish in a well because fishing, even if successful, takes time and money that is better spent drilling the well. Sometimes fishing attempts fail and part of or all the drilled well is lost.

Over the years, specialists known as *fishermen* have developed many ingenious tools and techniques to retrieve fish. One such tool is an *overshot*, which is a tube just large enough to slide over the top of the fish (fig. 216).

Inside the overshot are *grapples* that firmly latch onto the fish. The bottom of the overshot has a cut lip or *mule shoe* that acts as a guide and allows the overshot to be rotated over the fish. The overshot is run on drill pipe as part of a BHA that includes drill collars and *hydraulic jars*. When the overshot is slid over the top of the fish and the grapples are engaged, the crew pulls the overshot and the attached fish out of the hole.

Hydraulic jars are tools that allow a jarring blow to be delivered downhole on top of the fish. When the fish has been caught in the grapple, hydraulic jars apply force by cocking the jars to pressurize hydraulic fluid in the jar's chamber. The hydraulic pressure drives an internal piston that strikes an anvil inside the jars, similar to a hammer blow. The blow can be directed upwards or downwards as required for the particular fishing application. The force of the blow is quite large and is useful in freeing a fish stuck in the hole.

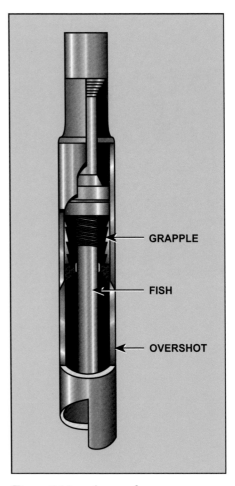

Figure 216.   An overshot

A *spear* (fig. 217) is a fishing tool that slides inside the fish and grips the fish from the inside. The spear is often run with jars.

Magnets and baskets are used to retrieve smaller pieces of junk left in a well. Although it is not a quick option, it is sometimes possible to drill small pieces of junk using drag bits with tungsten carbide placed on the bottom. These drag bits are called *junk mills* and are also used to smooth or *dress off* the top of a fish too large or jagged for an overshot to slide over.

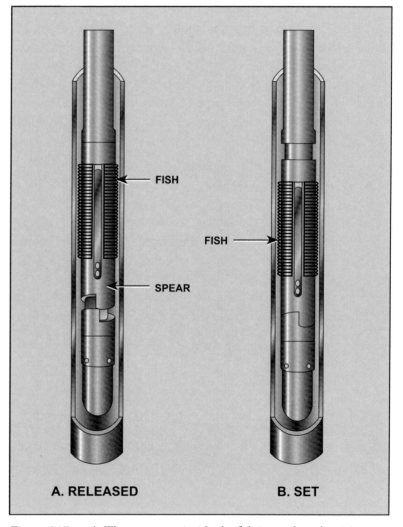

Figure 217. *A. The spear goes inside the fish in a released position. B. Once in position, the spear is set and the fish is removed.*

WELL CONTROL

A well is out of control if reservoir fluids are flowing in a way the operator or driller cannot regulate or stop. The most spectacular and hazardous type of uncontrolled well is the surface *blowout*. A surface blowout occurs when reservoir fluids escape from the well at the surface. Surface blowouts waste a valuable resource, can harm the long-term ability of the reservoir to produce, destroy the expensive drilling rig, and more importantly, hurt or kill people (fig. 218).

Uncontrolled flows can also occur in the well. *Downhole blowouts* happen when a high-pressure reservoir is allowed to flow fluids into a lower pressure reservoir at a different well depth. A downhole blowout does not produce escaping fluid at the surface. Nonetheless, it wastes hydrocarbons by flowing them into other reservoirs where they might become unrecoverable, damaging the ability of the flowing reservoir to sustain long-term flow. A downhole blowout can possibly charge the lower pressure zone with high pressure and create a future drilling hazard in the area.

Figure 218.   Fluids erupting from underground caught fire and melted this rig.

Blowouts occur through some form of human error. The reservoir pressures encountered when drilling the well might have been underestimated. The weight of the drilling mud might have been improperly controlled, or the drilling crew could have missed or misinterpreted signs of an impending blowout. The crew might have operated the blowout preventers improperly or the BOP equipment failed due to poor maintenance or incorrect design or installation.

Blowouts are prevented by keeping the pressure inside the wellbore higher than the pressures in the reservoirs being drilled. This is done by adjusting the mud weight. The mud weight is increased as required by adding barium sulfate (barite) to the mud. It is the responsibility of the drilling engineer and the mud engineer to predict the necessary mud weight. It is the derrickman's responsibility to make sure the mud is the required weight. If the drilling engineer or mud engineer fail to predict the correct reservoir pressures to be encountered, the mud weight might be too low. If the derrickman does not mix enough barite, the mud weight will also be too low. Either of these conditions causes the pressure in the wellbore to be less than the reservoir pressures encountered. If this occurs, the reservoir begins to flow fluids into the borehole.

If a reservoir begins to flow fluids into the borehole, the well is said to be kicking. Although the well has not yet formed a blowout, it will if corrective action is not taken. The drilling crews are trained to look for signs of a *kick*. The signs include an increasing level of mud in the active *mud pits* as reservoir fluids enter the borehole and push the mud out of the well. If the pumps are turned off during a kick, the well will actually be flowing on its own. These and other symptoms are sure signs of a kick.

When the drilling crew recognizes signs of a kick, the driller turns off the pumps, stops rotating the drill string, and pulls the drill string off the bottom, exposing the first tool joint above the slips. Then, one or more of the BOPs are closed so the flow is stopped at the surface. The BOPs are a series of valves that have been nippled up on the wellhead (fig. 219).

Several types of BOPs might be used. The top BOP is always an *annular preventer* that can close and seal around any shape of pipe protruding through the preventer. The annular preventer can even close over an open hole, if necessary. The preventers below the annular are a combination of pipe and *blind rams* (fig. 220). *Pipe rams* can close and seal only around a piece of pipe that matches the size of the hole machined into the ram. Blind rams can close and seal over an open hole but will not close around a piece of pipe. *Shear rams* can also be used; they are a special type of blind ram that can cut any pipe in the ram and then, seal over the hole created after severing the pipe.

Figure 219.   A stack of BOPs installed on top of the well

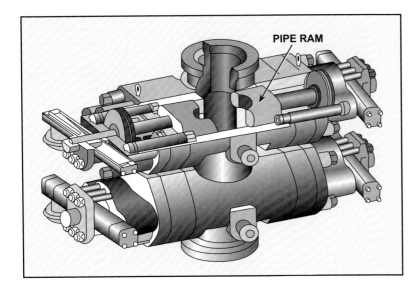

Figure 220.   Ram cutaway

Figure 221.   A subsea stack of BOPs being lowered to the seafloor from a floating rig

Onshore or on bottom-supported offshore rigs, the BOPs are installed beneath the floor of the rig. They can be operated manually or hydraulically. On floating offshore rigs, the BOPS are installed on the top of the wellhead on the ocean floor (fig. 221). They are operated either hydraulically or electrically. Subsea BOPs are connected back to the floating rig by a marine riser.

Once the BOPs are closed, the crew carefully monitors the shut-in pressure of the well while preparing to circulate the kicking reservoir fluids out of the well. If the shut-in pressure is allowed to grow, a lower pressure (weaker) reservoir in the well could fracture, allowing the kicking reservoir to flow into the lower pressured zone. This is the start of a downhole blowout.

The drilling crew circulates the reservoir fluids out of the well. The crew also circulates the old mud out of the well. It is replaced with new mud with a high enough mud weight to create a wellbore pressure higher than the pressure in the kicking reservoir. The circulation procedure is accomplished using the rig pumps and a *choke manifold* to control the escape of reservoir fluids and mud from the well (figs. 222, 223, and 224).

Figure 222.   Several valves and fittings make up a typical choke manifold.

Figure 223.   A remote-controlled choke installed in the choke manifold

Figure 224.   This control panel allows an operator to adjust the size of the choke.

After all the reservoir fluids and old mud are replaced by the new, heavier mud, the pressure in the well is again higher than the reservoir pressure. The reservoir can no longer flow into the well, and the well is said to be dead, or static. The BOPs can be opened, the mud conditioned, and drilling can safely resume.

It is important to note that well blowouts are reasonably rare. There are thousands of wells drilled without incident worldwide with few becoming well control problems or blowouts.

Safety training is part of everyday life for all hands on a drilling rig. There are safety meetings at the beginning of every tour and before each new part of a job. Outside training, such as well control schools and helicopter safety training for offshore crews, is also required for drilling personnel. The equipment used to drill a well is technical and complex, and those who run the equipment must be well trained.

The International Association of Drilling Contractors (IADC) keeps a list of detailed statistics on accident rates in the drilling industry. The annual statistics can be viewed at IADC's Web site (www.iadc. org). One statistic is the total number of any type of accident that occurs for every one million man-hours worked. This number has declined slightly from 2002–2007 at an average of 11.16 accidents per one million man-hours worked. A normal number of hours one person might work on a drilling rig is about 3,100 man-hours in one year. So, a five-person crew will work roughly 15,500 man-hours in one year. Using the average accident rate shown above, the number of accidents that might be estimated to occur in one year in the five-person crew is 0.18 or less than one. This calculation suggests the average drilling crewmember is a safe worker.

The total number of man-hours actually worked in the worldwide drilling industry is enormous and increases as more rigs are added to the world fleet. Table 3 shows the annual man-hours worked and the total accident frequency rate from the IADC data base. Although safety can always be improved, these statistics suggest the industry is becoming safer because the accident rate has decreased with the increasing work time.

Table 3
IADC Annual Work Time and Accident Statistics

Year	Total Man–Hours	Accident Frequency	Total Accidents
2007	446,335,455	10.24	4,572
2006	418,954,216	10.85	4,547
2005	369,693,317	11.72	4,332
2004	336,122,663	11.29	3,794
2003	301,959,960	11.16	3,369
2002	281,350,992	11.72	3,297

Source: IADC

Great care and planning goes into designing drilling operations to limit or eliminate environmental impact. Federal and state governments require a significant number of permits before operators can drill and produce hydrocarbons. Government regulations limit the amount of emissions from drilling and production sites. Contingency plans must be approved for spills and the possible occurrence of poison (H_2S) gas. Permits must be obtained for new well sites, offshore structures, and dredging operations. The discharge of wastewater and cuttings must be permitted and are monitored. Operators must also demonstrate the financial capability to take care of plugging and abandoning wells, cleaning up spills, restoring land sites, and decommissioning offshore structures when projects come to an end.

15

Conclusion

Drilling has developed into a specialized and technologically advanced business. The size of the equipment is enormous. The technical challenges to overcome as wells become deeper and are drilled in increasingly hostile environments are equally enormous. The technology of the most advanced drilling rig is computer-controlled and can be monitored from any office in the world. The guidance systems used in directional drilling rival those found on modern jet aircraft or spacecraft.

The energy business is the largest business in the world. This will continue because the standard of living in most countries is now tied to the ability to find and use energy efficiently. Well drilling continues to be an important part of the efficient use of energy, regardless of whether the well is producing hydrocarbons or water, or permanently disposing wastes by injecting them into deep layers in the earth.

The drilling industry must have people who are trained, motivated, and, most importantly, interested in the business, the science, and the art of drilling.

Appendix 1
Units of Conversion

Length	equal to	equal to	equal to
1 inch	0.025400 metres	2.540000 centimetres	25.40 millimetres
$^{1}/_{32}$nd of an inch	0.793750 millimetres		
1 foot	0.3048 metres	0.000189 miles	
1 yard	0.914400 metres		
1 mile	1.609344 kilometres	1609.34 metres	

Weight	equal to	equal to
1 pound-force	4.450000 newtons	0.445000 decanewtons

Pressure, Yield Point, and Gel Strength	equal to	equal to	equal to
1 pound/square inch	6895.00 Pa	6.895000 kPa	0.006895 megaPa
1 pound/square foot	47.882000 Pa	0.047882 kPa	0.000048 megaPa
1 pound/100 square feet	0.478820 Pa		

Volume	equal to	equal to	equal to	equal to
1 barrel	0.159000 cubic metres	159.00 litres	42.00 gallons	5.615 cubic feet
1 gallon	0.003785 cubic metres	3.79 litres	0.13369 cubic feet	
1 quart	0.946400 litres	0.25 gallons	0.033422 cubic feet	
1 cubic foot	0.028317 cubic metres	28.32 litres	7.48 gallons	
1 cubic inch	16.387000 cubic centimetres			
1 fluid ounce	29.570000 millilitres			
1 barrel per ton	0.175000 cubic metre/tonne			
1 pound per barrel	2.853000 kilogram/cubic metre			

Torque and Power	equal to
1 ft.lb	1.3558 newton metres
1 hp	0.746 kW

Viscosity	equal to
1 cp	0.001 Pa-sce

Appendix 2
Figures Credits

Figure		Owner	Web site
1.	Drilling rigs are large to accommodate the size of the drilling equipment and pipes.	PETEX	www.utexas.edu/ce/petex
2.	ConocoPhillips Britannia platform in the North Sea	Copyright © ConocoPhillips. All rights reserved.	www.conocophillips.com
3.	Drilling rig with a mast height of 147 feet (45 metres)	PETEX	www.utexas.edu/ce/petex
4.	Personal protective equipment (PPE) includes hard hats, gloves, hearing protection, and safety glasses.	PETEX	www.utexas.edu/ce/petex
5.	Steel stairways with handrails are used to get to the drilling rig floor.	PETEX	www.utexas.edu/ce/petex
6.	The drawworks is part of the hoisting system used to lift drill pipe into place.	PETEX	www.utexas.edu/ce/petex
7.	Whaling ships in New Bedford, Massachusetts	Source: National Oceanic and Atmospheric Administration, Department of Commerce	www.noaa.com
8.	Oilwells in Balakhani, a suburb of Baku, Azerbaijan in the late 1800s	Copyright © Brita Åsbrink Collection. All rights reserved.	
9.	Oil Creek near Titusville, Pennsylvania as it looks today	PETEX	www.utexas.edu/ce/petex
10.	Edwin L. Drake and his good friend Peter Wilson, a Titusville pharmacist, in front of the historic Drake well in 1861.	Source: Pennsylvania Department of Conservation and Natural Resources	
11.	Patillo Higgins	Public domain	
12.	Anthony Lucas, mining engineer at Spindletop	Public domain	
13.	Wall cake stabilizes the drilling hole.	PETEX	www.utexas.edu/ce/petex
14a.	The 1901 Lucas well is estimated to have flowed about 2 million gallons (7,570 cubic metres) of oil per day.	Public domain	
14b.	Spindletop oil field in 1903, two years after the first well was drilled	Public domain	
15.	A cable-tool rig	PETEX	www.utexas.edu/ce/petex
16.	A polycrystalline diamond compact bit (PDC) and a tri-cone bit	Copyright © Baker Hughes. All rights reserved.	www.bakerhughesdirect.com

Figure	Owner	Web site
17. The drill stem puts the bit on the bottom of the drilling hole.	PETEX	www.utexas.edu/ce/petex
18. Two floorhands place a joint of drill pipe in the mousehole prior to adding it to the active drill string.	PETEX	www.utexas.edu/ce/petex
19. Components in the rotary table rotate the drill string and bit.	PETEX	www.utexas.edu/ce/petex
20. A powerful motor in the top drive rotates the drill string and bit.	PETEX	www.utexas.edu/ce/petex
21. The bit is rotated by a downhole motor placed near it.	PETEX	www.utexas.edu/ce/petex
22. A pump circulates drilling mud down the drill pipe, out the bit, and up the hole.	PETEX	www.utexas.edu/ce/petex
23. Two pumps are available on this rig to move drilling fluid down the pipe.	PETEX	www.utexas.edu/ce/petex
24. Drilling mud	Copyright © Cobalt Environmental, Inc. All rights reserved.	www.cobalttanks.ca
25. A land rig	PETEX	www.utexas.edu/ce/petex
26. An offshore jackup rig	PETEX	www.utexas.edu/ce/petex
27. An inland barge rig	PETEX	
28. Rigs can be disassembled and moved piece-by-piece to a new location.	PETEX	www.utexas.edu/ce/petex
29. Types of MODUs	Public domain	
30. The first MODU was a posted-barge submersible designed to drill in shallow water.	PETEX	www.utexas.edu/ce/petex
31. When the bottles are flooded, the weight makes the bottle-type rig sink to the seafloor.	PETEX	www.utexas.edu/ce/petex
32. Ice floes on the North Bering Sea	Source: National Oceanic and Atmospheric Administration, Department of Commerce. Photograph by Commander Richard Behn, NOAA Corps.	www.noaa.com
33. A concrete island drilling system (CIDS) features a reinforced concrete caisson.	PETEX	www.utexas.edu/ce/petex
34. Drilling equipment is placed on the deck of a barge to drill in the shallow waters of bays and estuaries.	PETEX	www.utexas.edu/ce/petex
35. Four boats tow a jackup rig to its drilling location.	PETEX	www.utexas.edu/ce/petex
36. A jackup rig with four column-type legs	PETEX	www.utexas.edu/ce/petex

Figure	Owner	Web site
37. A jackup with open-truss legs	PETEX	www.utexas.edu/ce/petex
38. The hulls of these jackups are raised to clear the highest anticipated waves.	PETEX	www.utexas.edu/ce/petex
39. A semisubmersible rig floats on pontoons.	PETEX	www.utexas.edu/ce/petex
40. The heavy lift vessel, Blue Marlin, transporting BP's semisubmersible, Thunder Horse	Copyright © BP America Inc. All rights reserved.	www.bp.com
41. The pontoons of this semisubmersible float a few feet (metres) below the water's surface.	PETEX	www.utexas.edu/ce/petex
42. The main deck of a semisubmersible is huge. Shown here is the deck of the BP Thunder Horse.	Copyright © BP America Inc. All rights reserved.	www.bp.com
43. Deepwater Pathfinder 10,000-foot ultradeepwater drillship	Copyright © Transocean. All rights reserved.	www.deepwater.com
44. Marine riser	PETEX	www.utexas.edu/ce/petex
45a. The heave compensator keeps proper tension on the drill string.	PETEX	www.utexas.edu/ce/petex
45b. Heave compensator	Copyright © 2003-2007 IODP-USIO. All rights reserved.	www.iodp-usio.org
46. Workers on a drilling rig	Copyright © EnerMax, Inc. Photograph by Bret Boteler. All rights reserved.	www.enermaxinc.com
47. U.S. Department of the Interior, Minerals Management Service map of proposed sale of government mineral leases in 2001	Source: U.S. Department of the Interior, Minerals Management Service	www.doi.gov
48. IADC standard drilling bid form	Copyright © International Association of Drilling Contractors (IADC). All rights reserved.	www.iadc.org
49. A computer display showing a well log	PETEX	www.utexas.edu/ce/petex
50. A member of a casing crew stabs one joint of casing into another.	PETEX	www.utexas.edu/ce/petex
51. Personnel on this offshore rig enjoy quality food in the galley.	PETEX	www.utexas.edu/ce/petex
52. A driller on an offshore rig works in an environmentally controlled cabin.	PETEX	www.utexas.edu/ce/petex
53. The view from above the derrickman's position on the monkeyboard	PETEX	www.utexas.edu/ce/petex
54. A derrickman checking the weight or density of the drilling mud	PETEX	www.utexas.edu/ce/petex
55. Floorhands latch big wrenches called tongs onto the drill pipe.	PETEX	www.utexas.edu/ce/petex

Figure	Owner	Web site
56. Floorhands using power tongs to tighten drill pipe	PETEX	www.utexas.edu/ce/petex
57. Roustabouts move casing from a supply boat to the rig.	PETEX	www.utexas.edu/ce/petex
58. A crane operator manipulates controls from a position inside the crane cab.	PETEX	www.utexas.edu/ce/petex
59. A barge engineer monitors a semisubmersible's stability from a work station on board the rig.	PETEX	www.utexas.edu/ce/petex
60. BP's Thunder Horse listing in the Gulf of Mexico after a storm	Copyright © BP America Inc. All rights reserved.	www.bp.com
61. Arctic Discoverer LNG transport ship	Copyright © StatoilHydro. Photograph by Roar Lindefjeld. All rights reserved.	www.statoilhydro.com
62. A pore is a small open space in a rock.	PETEX	www.utexas.edu/ce/petex
63. A cross-section showing pore space and the small connections between larger pores	Source: National Energy Technology Laboratory, U.S. Department of Energy	www.netl.doe.gov
64. Connected pores give rocks permeability.	PETEX	www.utexas.edu/ce/petex
65. A fault trap and an anticlinal trap	PETEX	www.utexas.edu/ce/petex
66. Types of stratigraphic traps	PETEX	www.utexas.edu/ce/petex
67. A combination trap	PETEX	www.utexas.edu/ce/petex
68. A piercement salt dome	PETEX	www.utexas.edu/ce/petex
69. To the right of the tire, a large, heavy plate vibrates against the ground to create sound waves.	Copyright © Tethys Oil AB. All rights reserved.	www.tethysoil.com
70. Several special trucks vibrate plates against the ground.	Copyright © Tethys Oil AB. All rights reserved.	www.tethysoil.com
71. Fugro Explorer seismic vessel	Copyright © Fugro N.V. All rights reserved.	www.fugro.com
72. Stuck into the ground, a geophone picks up reflected sound waves.	Copyright © Dr. Lee M. Liberty. Boise State University. All rights reserved.	
73. iZone Virtual Reality room at EPI Centre in Rijswijk, the Netherlands, 2008.	Copyright © Photographic Services, Shell International Ltd. All rights reserved.	www.shell.com
74. Geologists working at a prospective petroleum area at the Peel Plateau in the Yukon	Copyright © Government of Yukon. Photograph by Tiffani Fraser. All rights reserved.	www.geology.gov.yk.ca
75. A reserve pit	PETEX	www.utexas.edu/ce/petex
76. Typical onshore layout of a drilling location	Dr. Paul Bommer	

Figure	Owner	Web site
77a. Pit cleaning with Super Vac units	Copyright © Northern A-1. All rights reserved	www.northerna1.com
77b. Reserve pit cleanup and removal	Copyright © Northern A-1. All rights reserved.	www.northerna1.com
78. A concrete pad to support the substructure surrounds this cellar.	PETEX	www.utexas.edu/ce/petex
79. The kelly has been placed in the rathole when the rig is not drilling.	PETEX	www.utexas.edu/ce/petex
80. A joint of drill pipe rests in this rig's mousehole.	PETEX	www.utexas.edu/ce/petex
81. A rathole rig drills the first part of the hole.	PETEX	www.utexas.edu/ce/petex
82. The conductor hole	PETEX	www.utexas.edu/ce/petex
83. The large-diameter pipe to the right is the top of the conductor pipe.	PETEX	www.utexas.edu/ce/petex
84. A portable shallow oil drilling rig	Copyright © Barbour Corporation. All rights reserved.	www.barbourwell.com
85a. A heavy lift vessel carries a semisubmersible to a new drilling location.	PETEX	www.utexas.edu/ce/petex
85b. The Black Marlin heavy lift vessel transporting the Nautilus rig	Copyright © Dockwise, Ltd. All rights reserved.	www.dockwise.com
86. A box-on-box substructure	PETEX	www.utexas.edu/ce/petex
87. A slingshot substructure is shown in folded position prior to being raised.	PETEX	www.utexas.edu/ce/petex
88. The slingshot substructure near its full height	PETEX	www.utexas.edu/ce/petex
89. This drawworks will be installed on the rig floor.	PETEX	www.utexas.edu/ce/petex
90. The drilling line is spooled onto the drawworks drum.	PETEX	www.utexas.edu/ce/petex
91. A mast being raised to a vertical position	PETEX	www.utexas.edu/ce/petex
92. This rig with a standard derrick was photographed in the 1970s at work in in West Texas.	PETEX	www.utexas.edu/ce/petex
93. The derrick supports the weight of the drill string and allows the drill string to be raised and lowered.	Copyright © Transocean. All rights reserved.	www.deepwater.com
94. The doghouse is located at the rig floor level.	PETEX	www.utexas.edu/ce/petex
95. In the foreground is a coal-fired boiler that made steam to power the cable-tool rig in the background.	PETEX	www.utexas.edu/ce/petex

Figure	Owner	Web site
96. A mechanical rig drilling in West Texas in the 1960s. Rigs like this still drill in many parts of the world today.	PETEX	www.utexas.edu/ce/petex
97. Three diesel engines power this rig.	PETEX	www.utexas.edu/ce/petex
98. Three engines drive a chain-and-sprocket compound to power equipment.	PETEX	www.utexas.edu/ce/petex
99. The diesel engine at right directly drives an alternating current electric generator.	PETEX	www.utexas.edu/ce/petex
100. Controls in the SCR house where AC electricity is converted to the correct DC voltage for the many DC motors powering this rig.	PETEX	www.utexas.edu/ce/petex
101. A motor-driven drawworks	PETEX	www.utexas.edu/ce/petex
102. Two powerful electric DC traction motors drive the drawworks on this rig.	PETEX	www.utexas.edu/ce/petex
103. The hoisting system	PETEX	www.utexas.edu/ce/petex
104. The drawworks	PETEX	www.utexas.edu/ce/petex
105. Removing the drawworks housing reveals the main brake bands to the left and right on the hubs of the drawworks drum.	PETEX	www.utexas.edu/ce/petex
106. The electromagnetic brake is mounted on the end of the drawworks.	PETEX	www.utexas.edu/ce/petex
107. A floorhand has a fiber rope wrapped around a friction cathead to lift an object on the rig floor.	PETEX	www.utexas.edu/ce/petex
108. Floorhand using an air hoist to lift an object	PETEX	www.utexas.edu/ce/petex
109. This makeup cathead has a chain coming out of it that is connected to the tongs.	PETEX	www.utexas.edu/ce/petex
110. Wire-rope drilling line coming off the drawworks drum	PETEX	www.utexas.edu/ce/petex
111. Drilling line is stored on this supply reel at the rig.	PETEX	www.utexas.edu/ce/petex
112. Drilling line is firmly clamped to this deadline anchor.	PETEX	
113. The sheaves (pulleys) of this crown block are near the bottom of the photo.	PETEX	www.utexas.edu/ce/petex

Figure		Owner	Web site
114.	Ten lines are strung between the traveling block and the crown block.	PETEX	www.utexas.edu/ce/petex
115.	Several wraps of drilling line on the drawworks drum	PETEX	www.utexas.edu/ce/petex
116.	Traveling block and kelly assembly	PETEX	www.utexas.edu/ce/petex
117.	The mast supports the blocks and other drilling tools.	PETEX	www.utexas.edu/ce/petex
118.	A rotary-table system	PETEX	www.utexas.edu/ce/petex
119.	The turntable is housed in a steel case.	PETEX	www.utexas.edu/ce/petex
120.	The master bushing fits inside the turntable.	PETEX	www.utexas.edu/ce/petex
121.	Crewmembers are installing one of two halves that make up the tapered bowl.	PETEX	www.utexas.edu/ce/petex
122.	Crewmembers set slips around the drill pipe and inside the master bushing's tapered bowl to suspend the pipe.	PETEX	www.utexas.edu/ce/petex
123.	The master bushing has four drive holes into which steel pins fit on the kelly drive bushing.	PETEX	www.utexas.edu/ce/petex
124.	A master bushing with a square bottom that fits into a square opening in the master bushing	PETEX	www.utexas.edu/ce/petex
125a.	A square kelly	PETEX	www.utexas.edu/ce/petex
125b.	A hexagonal kelly	PETEX	www.utexas.edu/ce/petex
126.	A hexagonal kelly inside a matching opening in the top of the kelly drive bushing	PETEX	www.utexas.edu/ce/petex
127.	The hook on the bottom of the traveling block is about to be latched onto the bail of the swivel.	PETEX	www.utexas.edu/ce/petex
128.	Drilling fluid goes through the rotary hose and enters the swivel through the gooseneck.	PETEX	www.utexas.edu/ce/petex
129.	A top drive, or power swivel, hangs from the traveling block and hook.	PETEX	www.utexas.edu/ce/petex
130.	Mud pressure pumped through the drill string forces the spiral rotor of the mud motor to turn inside the rubber helical-shaped stator.	PETEX	www.utexas.edu/ce/petex
131.	Horizontal hole	PETEX	www.utexas.edu/ce/petex

Figure	Owner	Web site
132. A downhole motor lying on the rack prior to being run into the hole	PETEX	www.utexas.edu/ce/petex
133. An adjustable bent housing on the motor deflects the bit a few degrees off-vertical to start the directional hole.	PETEX	www.utexas.edu/ce/petex
134. Drill collars are placed on the pipe rack prior to being run in the hole.	PETEX	www.utexas.edu/ce/petex
135. Drill collars put weight on the bit, which forces the bit cutters into the formation to drill it.	PETEX	www.utexas.edu/ce/petex
136. Several joints of drill pipe are placed on the pipe rack before being run in the well.	PETEX	www.utexas.edu/ce/petex
137. A floorhand stabs the pin of a joint of drill pipe into the box of another joint.	PETEX	www.utexas.edu/ce/petex
138. Two drill collars on a pipe rack; at left is the drill collar box; at right is the pin.	PETEX	www.utexas.edu/ce/petex
139. Drill collars racked in front of drill pipe on the rig floor	PETEX	www.utexas.edu/ce/petex
140. A roller cone bit has teeth (cutters) that roll, or turn, as the bit rotates.	PETEX	www.utexas.edu/ce/petex
141. Tungsten carbide inserts are tightly pressed into holes drilled into the bit cones.	PETEX	www.utexas.edu/ce/petex
142. Drilling fluid (salt water in this photo) is ejected out of the nozzles of a roller cone bit.	PETEX	www.utexas.edu/ce/petex
143. Bit cutaway showing internal bearing	Copyright © Halliburton. All rights reserved.	www.halliburton.com
144. Several types of natural diamond bits are available.	PETEX	www.utexas.edu/ce/petex
145. Several diamond-coated tungsten carbide disks (compacts) form the cutters on this polycrystalline diamond compact (PDC) bit.	PETEX	www.utexas.edu/ce/petex
146. Drilling mud swirls in one of several steel tanks on this rig.	PETEX	www.utexas.edu/ce/petex
147. A derrickman measures the density (weight) of a drilling mud sample using a balance calibrated in pounds per gallon.	PETEX	www.utexas.edu/ce/petex

Figure	Owner	Web site
148. Powerful mud pumps (most rigs have at least two) move drilling mud through the circulating system.	PETEX	www.utexas.edu/ce/petex
149. Components of a rig circulating system	PETEX	www.utexas.edu/ce/petex
150. The standpipe runs up one leg of the derrick, or mast, and conducts mud from the pump to the rotary hose.	PETEX	www.utexas.edu/ce/petex
151. Mud with cuttings falls over the vibrating shale shaker screen.	PETEX	www.utexas.edu/ce/petex
152. Desanders remove sand-sized particles from the mud.	PETEX	www.utexas.edu/ce/petex
153. Desilters remove smaller silt-sized particles from the mud.	PETEX	www.utexas.edu/ce/petex
154. The degasser removes a relatively small volume of gas that enters the mud from a downhole formation and is circulated to the surface in the annulus.	PETEX	www.utexas.edu/ce/petex
155. A centrifuge removes particles even smaller than silt.	PETEX	www.utexas.edu/ce/petex
156. A mud cleaner is used for mud weighted with barite.	PETEX	www.utexas.edu/ce/petex
157. Bulk barite tanks with bagged chemicals in the foreground	PETEX	www.utexas.edu/ce/petex
158. A derrickman, wearing personal protective equipment, adds dry components to the mud through a hopper.	PETEX	www.utexas.edu/ce/petex
159. A closed-top chemical barrel for adding caustic chemicals to the mud in the tanks	PETEX	www.utexas.edu/ce/petex
160. Typical wellbore architecture	Dr. Paul Bommer	
161. A bit being lowered into the hole on a drill collar	PETEX	www.utexas.edu/ce/petex
162. A kelly with related equipment in the rathole	PETEX	www.utexas.edu/ce/petex
163 Red-painted slips with three hand-grips suspend the drill string in the hole.	PETEX	www.utexas.edu/ce/petex
164 The kelly drive bushing is about to engage the master bushing on the rotary table.	PETEX	www.utexas.edu/ce/petex
165 The motor in the top drive turns the drill stem and the bit.	PETEX	www.utexas.edu/ce/petex

Figure	Owner	Web site
166. The black inner needle on the weight indicator shows the weight suspended from the derrick in thousands of pounds.	PETEX	www.utexas.edu/ce/petex
167. The kelly is drilled down (close to the kelly drive bushing), and it is time to to make a new connection.	PETEX	www.utexas.edu/ce/petex
168. Using the traveling block, the driller raises the kelly, exposing the first joint of drill pipe in the opening of the rotary table.	PETEX	www.utexas.edu/ce/petex
169. Crewmembers latch tongs on the kelly and on the drill pipe.	PETEX	www.utexas.edu/ce/petex
170. The kelly spinner rapidly rotates the kelly in or out of the drill pipe joint.	PETEX	www.utexas.edu/ce/petex
171. Crewmembers stab the kelly into the joint of pipe in the mousehole.	PETEX	www.utexas.edu/ce/petex
172. Crewmembers use tongs to buck up (tighten) one drill pipe joint to another.	PETEX	www.utexas.edu/ce/petex
173. Crewmembers remove the slips.	PETEX	www.utexas.edu/ce/petex
174. The kelly drive bushing is about to engage the master bushing.	PETEX	www.utexas.edu/ce/petex
175. Making a connection with a kelly	PETEX	www.utexas.edu/ce/petex
176. Making a connection using a top drive	PETEX	www.utexas.edu/ce/petex
177. An Iron Roughneck™ spins and bucks up joints with built-in equipment.	PETEX	www.utexas.edu/ce/petex
178. The kelly and swivel with its bail are put into the rathole.	PETEX	www.utexas.edu/ce/petex
179. Crewmembers latch elevators to the drill pipe tool joint suspended to the rotary table.	PETEX	www.utexas.edu/ce/petex
180. The floorhands set the lower end of the stand of pipe off to one side of the rig floor.	PETEX	www.utexas.edu/ce/petex
181. The derrickman places the upper end of a stand of drill pipe between the fingers of the fingerboard.	PETEX	www.utexas.edu/ce/petex
182. Making a trip	PETEX	www.utexas.edu/ce/petex
183. Top view of an automatic pipe handling device manipulating a stand of drill pipe.	PETEX	www.utexas.edu/ce/petex
184. A casing crewmember cleans and inspects the casing as it lies on the rack next to the rig.	PETEX	www.utexas.edu/ce/petex

Figure	Owner	Web site
185. Casing threads have been cleaned and inspected.	PETEX	www.utexas.edu/ce/petex
186. A joint of casing being lifted onto the rig floor	PETEX	www.utexas.edu/ce/petex
187. A joint of casing suspended in the mast; note the centralizer	PETEX	www.utexas.edu/ce/petex
188. Casing elevators suspend the casing joint as the driller lowers the joint into the casing slips.	PETEX	www.utexas.edu/ce/petex
189. Working from a platform called the stabbing board, a casing crewmember guides the casing elevators near the top of the casing joint.	PETEX	www.utexas.edu/ce/petex
190. Crewmembers lift the heavy steel-and-concrete guide shoe.	PETEX	www.utexas.edu/ce/petex
191. The guide shoe is made up on the bottom of the first joint of casing to go into the hole.	PETEX	www.utexas.edu/ce/petex
192. Cementing the casing: (A) the job in progress; (B) the finished job	PETEX	www.utexas.edu/ce/petex
193. Crewmembers install a float collar into the casing string.	PETEX	www.utexas.edu/ce/petex
194. Scratchers and centralizers are installed at various points in the casing string.	PETEX	www.utexas.edu/ce/petex
195. Top view of casing that is not centered in the hole	PETEX	www.utexas.edu/ce/petex
196. A cementing head (plug container) rests on the rig floor, ready to be made up on the last joint of casing to go into the hole.	PETEX	www.utexas.edu/ce/petex
197. To trip in, crewmembers stab a stand of drill pipe into another.	PETEX	www.utexas.edu/ce/petex
198. After stabbing the joint, crewmembers use a spinning wrench to thread the joints together.	PETEX	www.utexas.edu/ce/petex
199. After spin up, crewmembers use the tongs to buck up the tool joints to the correct torque.	PETEX	www.utexas.edu/ce/petex
200. A handful of cuttings made by the bit	PETEX	www.utexas.edu/ce/petex
201. Mud log section showing a formation that contains hydrocarbons	Dr. Paul Bommer	
202. Logging personnel run and control logging tools by means of wireline from a logging unit.	PETEX	www.utexas.edu/ce/petex

Figure	Owner	Web site
203. A well-site log is interpreted to give information about the formations drilled.	Dr. Paul Bommer	
204. Drill stem test tools	PETEX	www.utexas.edu/ce/petex
205. A successful DST	Stock photo	
206. Repeat formation tester (RFT) tool	Copyright © Schlumberger. All rights reserved.	www.slb.com
207a. A whole core barrel	PETEX	www.utexas.edu/ce/petex
207b. Sidewall coring device	PETEX	www.utexas.edu/ce/petex
208. A. An oil-saturated whole core from a South Texas well; B. Sidewall cores	PETEX	www.utexas.edu/ce/petex
209. This collection of valves and fittings is a Christmas tree.	PETEX	www.utexas.edu/ce/petex
210. Subsea wellheads	Copyright © GE Oil & Gas. All rights reserved.	www.geoilandgas.com
211. A coiled-tubing unit runs tubing into the well using a large reel.	PETEX	www.utexas.edu/ce/petex
212. Perforations (holes)	PETEX	www.utexas.edu/ce/petex
213. Shaped charges in a perforating gun make perforations.	PETEX	www.utexas.edu/ce/petex
214. A gravel pack	PETEX	www.utexas.edu/ce/petex
215. Several directional wells tap an offshore reservoir.	PETEX	www.utexas.edu/ce/petex
216. An overshot	PETEX	www.utexas.edu/ce/petex
217. A. The spear goes inside the fish in a released position. B. Once in position, the spear is set and the fish is removed.	PETEX	www.utexas.edu/ce/petex
218. Fluids erupting from underground caught fire and melted this rig.	PETEX	www.utexas.edu/ce/petex
219. A stack of BOPs installed on top of the well	PETEX	www.utexas.edu/ce/petex
220. Ram cutaway	PETEX	www.utexas.edu/ce/petex
221. A subsea stack of BOPs being lowered to the seafloor from a floating rig	PETEX	www.utexas.edu/ce/petex
222. Several valves and fittings make up a typical choke manifold.	PETEX	www.utexas.edu/ce/petex
223. A remote-controlled choke installed in the choke manifold	PETEX	www.utexas.edu/ce/petex
224. This control panel allows an operator to adjust the size of the choke	PETEX	www.utexas.edu/ce/petex

A

abandon *v*: to cease producing oil and gas from a well when it becomes unprofitable; to cease further work on a newly drilled well when it proves to contain unprofitable or no quantities of oil or gas. Also abandoning, abandonment.

AC *abbr*: alternating current.

accelerometer *n*: a device that reacts to changes in acceleration used in directional drilling as a way to determine the angle of the borehole with respect to vertical. As the hole takes on an off-vertical angle, the accelerometer detects the subtle change in the acceleration of gravity that will be proportional to the hole angle.

acid fracture *v*: to part or open fractures in limestone formations and etch pathways along the fractures using hydrochloric (HCl) acid under high pressure. See *formation fracturing*.

acidize *v*: to treat limestone or other formations with the appropriate acid for the purpose of increasing the rock permeability near the wellbore.

aft *n*: the back of a ship.

air drilling *n*: a method of rotary drilling that uses compressed air as the circulation medium. Natural gas can be substituted for air.

air hoist *n*: a hoist operated by compressed air; a pneumatic hoist. Air hoists are often mounted on the rig floor and are used to lift joints of pipe and other heavy objects.

alternating current (AC) *n*: electric current that reverses direction. The opposite of direct current (DC) which always flows in the same direction. AC current is widely used in power generation because it is more economical to transmit.

American Petroleum Institute (API) *n*: oil trade organization (founded in 1920) that is the leading standard-setting organization for all types of oilfield equipment. It maintains departments of production, transportation, refining, and marketing in Washington, DC. It offers publications regarding standards, recommended practices, and bulletins. www.api.org

anchor point *n*: a drilling line securing device that is fastened to a derrick or substructure leg. The anchor point must be strong enough to support the load carried by the deadline. See *deadline anchor*.

angle of inclination *n*: in directional drilling, the angle at which a well diverts from vertical; expressed in degrees, with vertical being 0°.

annular blowout preventer *n*: a large valve, usually installed above the ram preventers, that forms a seal in the annular space between the pipe and the wellbore or, if no pipe is present, in the wellbore itself. The seal is formed by a flexible element that is hydraulically forced to take on the shape of whatever is in the bore of the preventer. Compare *ram blowout preventer*.

annular preventer *n*: see *annular blowout preventer*.

annular space *n*: see *annulus*.

annulus *n*: the space between two concentric circles. In the petroleum industry, it is usually the space surrounding a pipe in the wellbore, or the space between tubing and casing, or the space between drill pipe and the wellbore.

anticlinal traps *n pl*: hyrdrocarbon structural traps closed in by an anticline or arch.

anticline *n*: rock layers folded in the shape of an arch. Anticlines sometimes trap oil and gas.

API *abbr*: *American Petroleum Institute*.

Arctic semisubmersible *n*: an offshore floating drilling vessel with a reinforced hull that will not be crushed by Arctic ice floes.

area drilling superintendent *n*: an employee of a drilling contractor whose job is to coordinate and oversee the contractor's drilling projects in a particular region or area. The drilling superintendent usually has responsibility for several drilling rigs.

artificial lift *n*: the use of artificial means to improve the flow of fluids from a production well.

assistant driller *n*: a member of a drilling rig crew whose job is to aid and assist the driller during rig operations. This person controls the drilling operation at certain times, keeps records, handles technical details, and, in general, keeps track of all phases of the operation. See *driller*.

assistant rig superintendent *n*: an employee of a drilling contractor whose job includes aiding the rig superintendent. The assistant rig superintendent may take over for the rig superintendent during nighttime hours. The assistant rig superintendent is sometimes called the night toolpusher. See *rig superintendent, toolpusher*.

automatic cathead *n*: a spool-shaped attachment on the catshaft. An automatic cathead is located on both sides of the drawworks on the catshaft. The driller can activate the cathead by pulling a control lever on the drawworks control panel. Automatic catheads are used to pull a chain to make up or break out joints of pipe. See *breakout cathead, friction cathead, makeup cathead*.

automatic pipe racker *n*: a device used on a drilling rig to automatically remove or insert drill stem components and store them temporarily in the fingerboard of the derrick. It replaces the need for a person to be in the derrick or mast when tripping pipe into or out of the hole.

B

background gas *n*: a small, measurable volume of natural gas carried in the drilling mud as the well is being drilled. The gas is not necessarily from any one sediment layer and is not considered an indicator of a commercial accumulation of natural gas. It is a base line value against which any changes in the natural gas content of the mud can be compared.

back off *v*: to unscrew one threaded piece (such as a section of pipe) from another.

back up *v*: to hold one section of an object such as pipe stationary while another section is being screwed into or out of it.

backup tongs *n*: a wrench that is placed on the drill string below the joint being added or removed. The wrench is secured to a derrick leg and prevents the drill string below the joint from turning.

bail *n*: a curved steel rod on top of the swivel that resembles the handle, or bail, of an ordinary bucket, but is much larger. The bail suspends the swivel from the hook on the traveling block. The two steel links that suspend the elevator from the hook are also called bails. *v*: to recover bottomhole fluids, samples, mud, sand, or drill cuttings by lowering a cylindrical vessel called a bailer to the bottom of a well, filling it, and retrieving it.

bailer *n*: a long, cylindrical container fitted with a valve at its lower end, used to remove water, sand, mud, drilling cuttings, or oil from a well in cable-tool drilling. It is also used on a slick line in well servicing operations.

barge *n*: a flat-decked, shallow-draft vessel, usually towed by a boat. A complete drilling rig may be assembled on a barge and the vessel used for drilling wells in lakes and in inland waters and marshes.

barge control operator *n*: an employee on a semisubmersible rig whose main duty is to monitor and control the stability of the rig. From a special work station on board the rig, this person controls the placement of ballast water inside the rig's pontoons to maintain the rig on even keel during all operations.

barge engineer *n*: see *barge control operator*.

barge master *n*: see *barge control operator*.

barite *n*: barium sulfate, $BaSO_4$; a dense mineral frequently used to increase the weight or density of drilling mud.

barrel (bbl) *n*: a measure of volume for petroleum products in the United States. One barrel is the equivalent of 42 U.S. gallons or 0.15899 cubic metres (9,702 cubic inches). One cubic metre equals 6.2897 barrels.

bbl *abbr*: barrel.

bed *n*: a specific layer of earth or rock that presents a contrast to other layers of different material lying above, below, or adjacent to it.

bedrock *n*: solid rock just beneath the soil.

belt *n*: a flexible band or cord connecting and wrapping around each of two or more pulleys to transmit power or impart motion.

belt guard *n*: a protective grill or cover for a belt and pulleys.

bent housing *n*: the bottom portion of a drilling motor, near the bit, that can be adjusted to a small angle. The use of a motor with a bend in the housing allows the well to be steered in a predetermined direction.

bent sub *n*: a short cylindrical device installed in the drill stem between the bottommost drill collar and a downhole motor. It deflects the downhole motor off vertical to drill a directional hole. Bent subs have been largely replaced by bent housing motors.

BHA *abbr*: *bottomhole assembly*.

BHL *abbr*: *bottomhole location*.

bit *n*: the cutting or boring tool used in drilling oil and gas wells. The bit consists of cutters and circulating ports or nozzles. The cutters can be steel teeth, tungsten carbide buttons, industrial diamonds, or polycrystalline diamond compacts (PDCs). Bits can be a roller cone style where three cones hold the cutters and roll on the bottom of the hole on bearings. Or, a bit can be a drag style where there are no moving parts and the cutters are dragged across the face of the rock by the motion of the drill string or downhole motor.

bit cutter *n*: the teeth or cutting structure of a bit.

bit sub *n*: a sub inserted between the drill collar and the bit. See *sub*.

blind ram *n*: an integral part of a blowout preventer that serves as the closing element on an open hole. Its ends do not fit around the drill pipe but seal against each other and shut off the space below completely. See *ram* and *shear ram*.

blind ram preventer *n*: a blowout preventer in which blind rams are the closing elements. See *blind ram*.

block *n*: one or more pulleys, or sheaves, mounted to rotate on a common axis. The crown block is an assembly of sheaves mounted on axles at the top of the derrick or mast. The drilling line is strung over the sheaves of the crown block alternately with the sheaves of the traveling block, which is raised and lowered in the derrick or mast by the drilling line.

blooie line *n*: a large diameter pipe used in air or gas drilling that conducts the air or gas returns flowing out of the well to a safe location away from the rig.

blowout *n*: an uncontrolled flow of gas, oil, or other well fluids into the atmosphere or into an underground formation. A blowout, or gusher, can occur when formation pressure exceeds the pressure applied to it by the column of drilling fluid.

blowout preventer (BOP) *n*: one of several valves installed at the wellhead to prevent the escape of fluids either in the annular space between the casing and the drill pipe or in open hole (i.e., hole with no drill pipe) during drilling or completion operations. See *annular blowout preventer, ram blowout preventer*.

blowout preventer control panel *n*: controls, usually located near the driller's position on the rig floor, that are manipulated to open and close the blowout preventers. See *blowout preventer*.

blowout preventer control unit *n*: a device that stores hydraulic fluid under pressure in special containers and provides a method to open and close the blowout preventers quickly and reliably. Hydraulic pressure from a pump provides the opening and closing force in the unit. See *blowout preventer*. Also called an accumulator.

BOP *abbr*: *blowout preventer*.

BOP stack *n*: the assembly of blowout preventers installed on a well.

bore *n*: the inside diameter of a pipe or a drilled hole. *v*: to penetrate or pierce with a rotary tool.

borehole *n*: a hole made by drilling or boring; a wellbore.

bottle type *n*: a submersible (bottom-supported) rig that has large cylinders (bottles) at the corners that can be flooded with seawater to cause the rig to sink to the bottom of the sea. These rigs have largely been replaced by jackup-style rigs. Some bottle-type submersibles have been converted to semisubmersible service.

bottomhole *n*: the lowest or deepest part of a well. *adj*: pertaining to the bottom of the wellbore.

bottomhole assembly (BHA) *n*: the portion of the drilling assembly below the drill pipe. It can be very simple—composed of only the bit and drill collars—or, it can be very complex and made up of several drilling tools.

bottomhole location (BHL) *n*: the location of the bottom of a well that has been deviated from vertical.

bottomhole pressure *n*: the pressure at the bottom of a borehole caused by the hydrostatic pressure of the wellbore fluid and, sometimes, by any back-pressure held at the surface, as when the well is shut in with blowout preventers. The bottomhole pressure may also refer to the pressure contained in a reservoir.

bottom plug *n*: a cement plug that precedes a cement slurry being pumped down the casing. The plug wipes drilling mud off the walls of the casing and prevents it from contaminating the cement. See *cementing*.

bottom-supported unit *n*: mobile offshore drilling units (MODU) that contact the ocean bottom and are supported by it.

box *n*: the threaded female section of a connection. See *tool joint*.

box and pin *n*: see *tool joint*.

box-on-box *adj*: a type of rig substructure using steel frame boxes to elevate the rig floor.

brake *n*: a device for arresting the motion of a mechanism, usually by means of friction, as in the drawworks brake.

brake band *n*: a part of the brake mechanism consisting of a flexible steel band lined with a material that grips a drum when tightened. On a drilling rig, the brake band acts on the flanges of the drawworks drum to control the speed of the traveling block and its load of drill pipe, casing, or tubing.

break *v*: to begin or start, as in to break tour.

break out *v*: to unscrew one section of pipe from another section, especially drill pipe while it is being withdrawn from the wellbore.

breakout cathead *n*: a device attached to the catshaft of the drawworks that is used as a power source for unscrewing drill pipe; it is the automatic cathead located opposite the driller's side of the drawworks. See *cathead*. Compare *makeup cathead*.

breakout tongs *n pl*: tongs that are used to start unscrewing one section of pipe from another section, especially drill pipe coming out of the hole. See *tongs*.

break tour *v*: to begin operating 24 hours a day.

brine *n*: a solution of salt and fresh water.

bring in a well *v*: see *complete the well*.

British thermal unit (Btu) *n*: a measure of the energy content of a substance. One Btu is defined as the amount of heat necessary to raise the temperature of one pound of fresh water one degree Fahrenheit. For example, the energy content of natural gas is often found to be around 1,000 Btu per standard cubic foot.

Btu *abbr*: *British thermal unit*.

buck up *v*: to tighten up a threaded connection (such as two joints of drill pipe).

bulk tank *n*: on a drilling rig, a large metal bin that usually holds a large amount of a certain mud additive, such as barite, that is used in large quantities in the makeup of the drilling fluid.

bullwheel *n*: one of several large wheels joined by an axle and used to hold the drilling line on a cable-tool rig.

bump *adj*: in cementing operations, pertaining to a cement plug that comes to rest on the float collar. A cementing operator may say, "I have a bumped plug" when the plug strikes the float collar. Bumping a plug results in a sudden increase in pump pressure and is the sure sign that the plug has reached the float collar.

bushing *n*: 1. a pipe fitting on which the external thread is larger than the internal thread to allow two pipes of different sizes to be connected. 2. a removable lining or sleeve inserted or screwed into an opening to limit its size, resist wear or corrosion, or serve as a guide. Definition no. 2 is the most obvious use of bushings on the drilling rig. See *kelly bushing*, *master bushing*.

C

cable *n*: a rope woven of wire, hemp, or other strong fibers. See *wire rope*.

cable-tool drilling *n*: a drilling method in which the hole is drilled by dropping a sharply pointed percussion bit on bottom. The bit is attached to a cable, and the cable is repeatedly raised and dropped as the hole is drilled.

cable-tool rig *n*: a drilling rig that uses wire-rope cable to suspend a weighted, chisel-shaped bit in the hole. Machinery on the rig repeatedly lifts and drops the cable and bit. Each time the bit strikes the bottom of the hole, it crushes rock and drills deeper. Rotary drilling rigs have virtually replaced all cable-tool rigs.

caisson *n*: one of several columns made of steel or concrete that serve as the foundation for a rigid offshore platform rig, such as the concrete gravity platform rig.

cap *n*: to control a well that is flowing out of control; often accomplished by attaching a valve in the open position on top of the well and then closing it to seal off the flow.

cased *adj*: pertaining to a wellbore in which casing has been run and cemented. See *casing*.

cased hole *n*: a wellbore in which casing has been run.

casing *n*: steel pipe placed in an oil or gas well to prevent the wall of the hole from caving in and, if cemented in place, to prevent movement of fluids from one formation to another.

casing centralizer *n*: a device secured around the casing to center it in the hole. Casing that is centralized allows a more uniform cement sheath to form around the pipe.

casing coupling *n*: a tubular section of pipe that is threaded inside and connects two joints of casing. The collar is not integral to the casing, but is separate. It is screwed on to one pin or male end of the casing with the other end of the collar exposed so that the pin end of the next joint can be made up in the top of the collar as the casing is run into the well.

casing crew *n*: the employees of a company that specializes in preparing and running casing into a well. The casing crew usually makes up the casing as it is lowered into the well. The regular drilling crew also assists the casing crew in its work.

casing elevator *n*: see *elevators*.

casing pipe *n*: see *casing*.

casing string *n*: the entire length of all the joints of casing run in a well.

cathead *n*: a spool-shaped attachment on the catshaft. An automatic and a friction cathead are located on both sides of the drawworks on the catshaft. Catheads are used to pull a rope or a chain to hoist objects or for use in making up or breaking out joints of pipe. See *automatic cathead, breakout cathead, friction cathead, makeup cathead*.

catline *n*: a hoisting or pulling line powered by the cathead and used to lift heavy equipment on the rig. See *cathead*.

catshaft *n*: an axle that crosses through the drawworks and contains a revolving spool called a cathead at either end. See *cathead*.

catwalk *n*: a platform on the side of the drilling rig used as a staging area for tools, pipe, and other drilling equipment.

caving *n*: collapsing of the walls of the wellbore. Also called caving in.

cellar *n*: a pit dug in the ground and lined with either concrete, wood, or corrugated metal, that provides additional room between the rig floor and the first element of the wellhead to accommodate the installation of the blowout preventers. This allows for the use of a wide variety of wellhead types without altering the height of the substructure.

cement *n*: a powder consisting of alumina, silica, lime, and other substances that hardens when mixed with water. Extensively used in the oil industry to bond casing to the walls of the wellbore and to seal unwanted portions of a well.

cementing *n*: the application of a liquid slurry of cement and water to various points inside or outside the casing.

cementing company *n*: a company whose specialty is preparing, transporting, and pumping cement into a well. A cementing company's crew usually pumps the cement to secure casing in the well.

cementing head *n*: an accessory attached to the top of the casing to facilitate cementing of the casing with passages for cement slurry and retainer chambers for cementing plugs. Also called a plug retainer.

cement plug *n*: a portion of cement placed at some point in the wellbore to seal it.

centralizer *n*: see *casing centralizer*.

centrifuge *n*: a device that spins in a fixed axis. If a suspension of particles is placed in the spinning device, centrifugal force will cause the denser particles to separate from the less dense particles.

choke *n*: a device with an orifice installed in a line to restrict the flow of fluids.

choke manifold *n*: an arrangement of piping and special valves, called chokes. In drilling, reservoir fluids and drilling mud are circulated through a choke manifold when the blowout preventers are closed. This allows the well to be kept under control.

Christmas tree *n*: the control valves, pressure gauges, and chokes assembled at the top of a well to control the flow of oil and gas after the well has been drilled and completed.

chromatographic analysis *n*: a laboratory process where the various components or chemicals that make up a substance can be separated and identified.

circulate *v*: to flow from a beginning point in a system and return to the starting point. For example, drilling fluid is pumped out of the suction pit, down the drill string, out the bit, up the annulus, and back to the pits while drilling proceeds. This process is called circulating drilling fluid.

circulating components *n pl*: the equipment included in the drilling fluid circulating system of a rotary rig. The basic components consist of the mud pump and surface lines, the stand pipe and rotary hose, the swivel (or top drive); the drill stem (to include the kelly if one is being used), the bit, and the mud return line.

circulation *n*: the movement of drilling fluid out of the mud pits, down the drill stem, up the annulus, and back to the mud pits.

clay *n*: a group of hydrous aluminum silicate minerals (clay minerals). When mixed with water these minerals hydrate to create viscosity and gel strength that improve the capacity of the drilling mud to suspend and remove drilled cuttings from the well. The hydrated clay particles also form the mud cake on the wall of the wellbore. The most common clay type used is bentonite, often referred to as gel.

coiled tubing *n*: a continuous string of flexible steel tubing, often hundreds or thousands of feet long wound onto a large diameter reel. The reel is an integral part of the coiled tubing unit, consisting of several devices that ensure the tubing can be safely inserted into the well from the surface. Because tubing can be lowered into a well without making up joints of tubing, running coiled tubing into the well is often faster and less expensive than running conventional tubing. Rapid advances in the use of coiled tubing make it a popular way in which to run tubing into and out of a well. Also called reeled tubing.

collar *n*: 1. a coupling device used to join two lengths of pipe, such as casing or tubing. 2. a drill collar. See *drill collar*.

combination trap *n*: 1. a subsurface hydrocarbon trap that has the features of both a structural trap and a stratigraphic trap. 2. a combination of two or more structural traps or two or more stratigraphic traps.

come out of the hole *v*: also called trip out.

compact *n*: a small disk made of tungsten carbide. See *insert*.

company man *n*: see *company representative*.

company representative *n*: an employee of an operating company whose job is to represent the company's interests at the drilling location.

complete the well *v*: to finish work on a well and bring it to productive status. See *well completion*.

completion rig *n*: a smaller rig used for well completion operations.

compound *n*: a mechanism used to transmit power from the engines to the pump, the drawworks, and other machinery on a drilling rig. It is composed of clutches, chains and sprockets, and a number of shafts, both driven and driving. *v*: to connect two or more power-producing devices, such as engines, to run driven equipment, such as the drawworks.

condensate *n*: moisture that has condensed or liquefied during the process of distillation.

condition *v*: to prepare or mix drilling mud.

condition mud *v*: a process whereby the drilling mud properties are adjusted to desired levels. It is done by circulating the mud through the mud system and adding whatever chemicals and liquids are necessary to adjust the properties of the mud.

conductor casing *n*: the first string of casing in a well used to prevent the soft formations near the surface from caving in and to conduct drilling mud from the bottom of the hole to the surface when drilling starts. Also called conductor pipe, drive pipe.

conductor hole *n*: the starting hole for the well. It contains the large diameter conductor casing that prevents the hole from caving in and conveys or conducts the drilling fluid back to the mud system while drilling the next section.

conductor pipe *n*: see *conductor casing*.

cone *n*: a conical-shaped metal device into which cutting teeth are formed or mounted on a roller cone bit. See *roller cone bit*.

confirmation well *n*: the second producer in a new field, following the discovery well. A confirmation well is placed at a strategic location designed to prove or disprove the size of a reservoir. There may be more than one confirmation well.

connection *n*: 1. the action of adding a joint of pipe to the drill stem as drilling progresses. 2. a section of pipe or a fitting used to join pipe to pipe or to a vessel.

contract *n*: a written agreement that can be enforced by law and that lists the terms under which the acts required are to be performed. A drilling contract covers such factors as the cost of drilling the well (whether turnkey, by the foot, or by the day), the distribution of expenses between operator and contractor, and the type of equipment to be used. *v*: to decrease in size.

core *n*: a cylindrical sample taken from a formation for geological analysis. *v*: to obtain a solid, cylindrical formation sample for analysis. See *sidewall core, whole core*.

core barrel *n*: a tubular device. For whole cores, it is usually from 10 to 60 feet (3 to 18 metres) long and is run at the bottom of the drill pipe with a core bit or head and is used to store and recover the core sample that is cut. For sidewall cores, the barrel is an empty tube about 2" x 3" and is fired into the side of the hole from a sidewall core gun. The sidewall core becomes lodged in the short barrel and is retrieved from the well.

crane *n*: a machine for raising, lowering, and revolving heavy pieces of equipment, especially on offshore rigs and platforms.

crane operator *n*: a person responsible for the use of the cranes on a rig. A member of the support crew on an offshore rig.

crew *n*: the workers on a drilling rig, including the driller, the derrickhand, and the floorhands. Also called crewmember.

crown *n*: the crown block or top of a derrick or mast.

crown block *n*: an assembly of sheaves mounted on axles at the top of the derrick or mast and over which the drilling line is strung. See *block*.

crude oil *n*: unrefined liquid petroleum. It ranges in density from very light to very heavy and in color from yellow to black, and it may have a paraffin, asphalt, or mixed base.

cutters *n pl*: cutting teeth or structure on a bit or reamer.

cuttings *n pl*: the fragments of rock broken by the bit and brought to the surface in the drilling fluid.

D

darcy *n*: the measure of permeability (square metres), named after the French engineer Henry Darcy.

daywork contract *n*: a type of drilling contract where the work performed under the contract is paid for by a certain amount each day.

DC *abbr*: *direct current.*

deadline *n*: the drilling line from the crown block sheave to the anchor that does not move.

deadline anchor *n*: see *deadline tie-down anchor.*

deadline tie-down anchor *n*: a device to which the deadline is attached and securely fastened to the mast or derrick substructure. Also called a deadline anchor.

degasser *n*: the device used to remove gas from drilling fluid.

density *n*: the mass or weight of a substance per unit volume. For drilling mud, the mud weight is commonly expressed in pounds/gallon or ppg.

density log *n*: a device used to measure the porosity of a formation. The device measures the response of a formation when it is bombarded with gamma rays. The response is proportional to the bulk density of the formation. The bulk density is made up of the amount of rock and fluid present in the formation. Therefore, the log is a measurement of the formation porosity if the type of rock and pore fluid are known. The gamma ray source and the detectors are mounted on a pad that must be placed in contact with the borehole wall. Due to the requirement that the pad be in contact with the wall of the hole, the density log is not used inside of casing.

derrick *n*: a large, load-bearing structure usually of bolted construction. In drilling, the standard derrick has four legs standing at the corners of the substructure and reaching to the crown block. Compare *mast.*

derrickman *n*: the crewmember who works in the derrick on the monkeyboard and handles the upper end of the drill string as it is being hoisted out of or lowered into the hole. This person is also responsible for the circulating machinery and the conditioning of the drilling fluid.

desander *n*: a centrifugal device for removing sand, down to 50 micron size, from drilling fluid to prevent abrasion of the pumps and control the solids content of the drilling mud. Compare *desilter.*

desilter *n*: a centrifugal device for removing very fine particles, or silt down to 20 micron size, from drilling fluid to keep the amount of solids in the fluid at the lowest possible point. Compare *desander.*

development well *n*: 1. a well drilled in a field in proven territory to complete a pattern of production. 2. an exploitation well.

diamond bit *n*: a drill bit that has small industrial diamonds embedded in the matrix or bit face. Cutting is performed by the rotation of the very hard diamonds over the rock surface.

diapir *n*: a dome or anticlinal fold in which a mobile plastic core has ruptured the more brittle overlying rock. Commonly made of salt, but it can also be shale. Also called piercement dome.

diesel engine *n*: a high-compression, internal-combustion engine used extensively for powering drilling rigs. In a diesel engine, air is drawn into the cylinders and compressed to very high pressures; ignition occurs as fuel is injected into the compressed and heated air. Combustion takes place within the cylinder above the piston, and expansion of the combustion products imparts power to the piston.

displacement fluid *n*: any fluid that is pumped after another fluid in order to move or displace the first fluid to a desired location in the well. For example, water is often used as the displacing fluid for cement.

direct current (DC) *n*: an electric current that moves only in one direction. DC current is often used to power the traction motors on an electric drilling rig. Compare *alternating current.*

directional drilling *n*: the intentional deviation of a wellbore from the vertical.

directional hole *n*: a wellbore intentionally drilled at an angle from the vertical.

displacement fluid *n*: the fluid, usually drilling mud or water, that is pumped into the well after cementing to force the cement out of the casing and into the annulus. The cement is displaced to the desired position in the well and the casing left full of a fluid that is useful for the next operation.

doghouse *n*: a small enclosure on the rig floor used as an office for the driller and as a storehouse for small objects.

dolomite *n*: a sedimentary carbonite rock often present in limestone that occurs in deep-sea sediments, where organic matter content is high.

double *n*: a length of drill pipe, casing, or tubing consisting of two joints screwed together. Compare *quadruple, single, triple.*

downhole *adj, adv*: pertaining to a location in the wellbore.

downhole blowout *n*: an unintended and uncontrolled flow of fluids from one reservoir into another. This flow does not reach the surface of the ground.

downhole motor *n*: a drilling tool comprised of a helical rotor and stator made up in the drill string directly above the bit. It causes the bit to turn because the rotor turns when drilling mud is pumped through the motor.

drag bit *n*: any bit with no moving parts. The cutting structure of the bit is dragged across the rock face by the motion of the drill string or downhole motor. Compare *roller cone bit*.

Drake well *n*: the first well drilled in the United States in search of oil. Some 69 feet (21 metres) deep, it was drilled near Titusville, Pennsylvania, and was completed in 1859. It was named after Edwin L. Drake, who was hired by the well owners to oversee the drilling.

drawworks *n*: the hoisting mechanism on a drilling rig. It is a large winch that spools off or takes in the drilling line and thus raises or lowers the traveling block, the drill stem and the bit.

drawworks brake *n*: the mechanical brake on the drawworks that can prevent the drawworks drum from moving.

drawworks drum *n*: the spool-shaped cylinder in the drawworks around which drilling line is wound, or spooled.

dress off *v*: to smooth off or polish. For example, the jagged top of a piece of drill pipe that has been lost in a well may be smoothed or dressed off by using a mill.

drill *v*: to bore a hole in the earth.

drill ahead *v*: to continue drilling operations.

drill bit *n*: the cutting or boring element used for drilling. See *bit*.

drill collar *n*: a heavy, thick-walled tube, usually steel, placed between the drill pipe and the bit in the drill stem. Several drill collars are used to apply weight on the bit.

drill collar sub *n*: a short piece of pipe (sub) used to match different thread types between drill collars and drill pipe.

driller *n*: the employee directly in charge of a drilling rig and crew whose main duty is operation of the drilling and hoisting equipment. The driller is also responsible for the condition of the well and the supervision of the drilling crew.

driller's console *n*: control equipment on the rig floor with which the driller operates the various components of the drilling rig.

driller's control panel *n*: see *driller's console*.

drilling company *n*: see *drilling contractor*.

drilling contract *n*: an agreement made between a drilling company and an operating company to drill and complete a well. It sets forth the obligation of each party, compensation, method of drilling, depth to be drilled, etc.

drilling contractor *n*: an individual or group who owns a drilling rig or rigs and contracts services for drilling wells.

drilling crew *n*: a driller, a derrickhand, and two or more floorhands who operate a drilling rig.

drilling engineer *n*: an engineer who specializes in the technical aspects of drilling.

drilling fluid *n*: circulating fluid, one function of which is to lift cuttings out of the wellbore and to the surface. Other functions are to cool and lubricate the bit and the drill string and to offset downhole formation pressures. Drilling fluids can be air, natural gas, foam, water, water and clay mud mixtures or diesel, or synthetic oil mixtures. See *mud*.

drilling line *n*: a wire rope used to support the drilling tools.

drilling mud *n*: a specially compounded liquid circulated through the wellbore during rotary drilling operations. See *drilling fluid, mud*.

drilling rate *n*: the speed with which the bit drills the formation; usually called the rate of penetration (ROP).

drill pipe *n*: jointed steel pipe made up in the drill stem between the kelly or top drive on the surface and the bottomhole assembly on the bottom. Joints are made up (screwed together) to form the drill string.

drillship *n*: a self-propelled floating offshore drilling unit that is a ship constructed to permit a well to be drilled from it. Although not as stable as semisubmersibles, drillships are capable of drilling exploratory wells in deep, remote waters. See *floating offshore drilling rig*.

drill site *n*: the location of a drilling rig.

drill stem *n*: all members in the assembly used for rotary drilling from the swivel to the bit, including the kelly, the drill pipe, the drill collars, the stabilizers, and various specialty items. Compare *drill string*.

drill stem test (DST) *n*: a method of formation testing. The basic drill stem test tool consists of a packer or packers, valves, or ports that may be opened and closed from the surface, and two or more pressure-recording devices. The tool is lowered on the drill string to the zone to be tested. The packer or packers are set to isolate the zone from the rest of the well. The valves or ports are then opened to allow for formation flow while the recorders chart pressures. A sampling chamber traps clean formation fluids at the end of the test.

drill string *n*: the column, or string, of drill pipe with attached tools that transmits fluid and rotational power from the kelly to the bit. Compare *drill stem*.

drive bushing *n*: see *kelly drive bushing*.

drive pipe *n*: see *conductor casing*.

drum *n*: a cylinder around which wire rope is wound in the drawworks.

DST *abbr*: *drill stem test*.

dynamic positioning *n*: a method by which a floating offshore drilling rig is maintained in position over an offshore well location without the use of mooring anchors. Several propulsion units, called thrusters, are located on the hulls of the structure and are actuated by a sensing system. A computer to which the system feeds signals directs the thrusters to maintain the rig on location.

dynamic positioning operator *n*: an employee on a drillship or semisubmersible drilling rig whose primary duty is to monitor, operate, and maintain the equipment that keeps the rig on station while drilling.

E

electric log *n*: an electrical survey made of certain electrical characteristics to identify the formations, determine the nature and amount of fluids contained, and estimate depth.

electric rig *n*: a drilling rig on which the energy from the power source—usually several diesel engines—is changed to AC electricity by generators mounted on the engines. The AC power is converted to the correct voltage of DC power in the control or SCR house. The correct voltage of DC current is sent from the SCR house to the various DC motors used to drive the rig components like the drawworks and the pumps. Compare *mechanical rig*.

elevators *n pl*: clamps that grip a joint of casing, tubing, drill collars, or drill pipe so that the joint can be raised from or lowered into the hole.

engine *n*: a machine for converting the heat content of fuel into rotary motion that can be used to power other machines.

exploration *n*: the search for reservoirs of oil and gas, including aerial and geophysical surveys, geological studies, core testing, and drilling of wildcats.

explorationists *n pl*: technically trained people, mainly geologists and geophysicists, who determine likely places where oil and gas may exist.

exploration well *n*: a well drilled either in search of an as-yet-undiscovered pool of oil or gas (a wildcat well) or to extend greatly the limits of a known pool (a step-out well).

F

fastline *n*: the end of the drilling line that is affixed to the drum or reel of the drawworks. It travels with greater velocity than any other portion of the line as it is spooled or unspooled from the drawworks drum. Compare *deadline*.

fault *n*: a break in the Earth's crust along which rocks on one side have been displaced (upward, downward, or laterally) relative to those on the other side.

faulted anticline *n*: a combined structural feature where a layer of rock has been folded into an anticline and then broken or faulted.

fault trap *n*: a subsurface hydrocarbon trap created by faulting in which an impermeable rock layer has moved opposite the reservoir bed and stopped fluid migration.

female connection *n*: a pipe, a coupling, or a tool threaded on the inside so that only a male connection can be threaded into it. Also called the box. Compare *male connection*.

field *n*: a geographical area in which a number of oil or gas wells produce from a continuous reservoir. A field may refer to surface area only or to underground productive formations as well. See *oilfield*.

fingerboard *n*: a rack that appears as a series of fingers with spaces in between the fingers. It supports the tops of the stands of pipe being stacked or "racked back" in the derrick or mast.

fish *n*: an object that is left in the wellbore during drilling or workover operations and that must be recovered before work can proceed. *v*: to recover from a well any equipment left there during drilling operations, such as a lost bit or drill collar or part of the drill string.

fishermen *n pl*: persons who have the skill and the equipment necessary to attempt to recover objects that have been lost in a well.

fishing *n*: the procedure of recovering lost or stuck equipment in the wellbore. See *fish*.

fishing jars *n*: see *hydraulic jars*.

fishing tool *n*: a tool designed to recover equipment lost in a well.

flapper *n*: a flow control device where a door or flapper will open to allow flow in one direction, but will close if the flow should attempt to reverse direction.

flare *v*: the act of burning a flow of hydrocarbons, such as a flowing gas stream.

float collar *n*: a special coupling device inserted one or two joints above the bottom of the casing string that contains a check valve to permit fluid to pass downward but not upward through the casing. The float collar prevents drilling mud from entering the casing while it is being lowered and it prevents backflow of cement during a cementing operation.

floater *n*: see *floating offshore drilling rig.*

floating offshore drilling rig *n*: a type of mobile offshore drilling unit that floats and is not in contact with the seafloor (except with anchors) when it is in the drilling mode. Floating units include drillships and semisubmersibles. See *mobile offshore drilling unit.*

floe *n*: a floating ice field of any size.

floorhand *n*: a worker on a drilling or workover rig, subordinate to the driller and the derrickhand, whose primary work station is on the rig floor. Also called a rotary helper, floorman, rig crewman, or roughneck.

flow diverter *n*: a device that directs the fluids flowing from a well to a safe location away from the rig. The device is similar to an annular blowout preventer.

fluid *n*: a substance that flows and yields to any force tending to change its shape. Liquids and gases are fluids.

footage rate *n*: a fee basis in drilling contracts stipulating that payment to the drilling contractor is made according to the number of feet or metres of hole drilled. See *metreage contract.*

fore *n*: the front of a ship.

forge *v*: to use hard blows to form and shape metallic ingots into useful items.

formation *n*: a bed or deposit composed throughout of substantially the same kind of rock. Each formation is given a name, frequently as a result of the study of the formation outcrop at the surface and sometimes based on fossils found in the formation.

formation evaluation *n*: a process where the rock type, the properties of the rock, such as the porosity, and the type and amount of fluids held in the porosity are determined. Formation evaluation can be done with well logs, cores, and flow tests.

formation fluid *n*: fluid (such as gas, oil, or water) that exists in a subsurface rock formation.

formation fracturing *n*: a method of stimulating production by opening new flow channels in the rock surrounding a production well. Often called a frac job. A fracturing fluid (such as water, oil or deisel or acid) is pumped into a formation at a high enough pressure so that the rock cracks open forming a fracture. The fracture can be extended by continued high pressure injection of the fracturing fluid into the formation. The fracture is filled with material to hold the fracture open when the pressure is removed. The material, called proppant, can be sand or any number of other materials. If the rock is a carbonate, like limestone, acid can be used to etch pathways along the fracture face.

When the pressure is released at the surface, the fracturing fluid returns to the well and the fracture closes partially on the proppant or etched channels. The propped fracture provides a high permeability pathway for reservoir fluids to use to flow to the wellbore. Also called fracturing.

frac job *n*: see *formation fracturing.*

friction cathead *n*: a spool-shaped attachment on the end of the catshaft around which a rope for hoisting and moving heavy equipment on or near the rig floor is wound by a crewmember.

G

gamma ray log *n*: a device that measures the amount of naturally occurring gamma ray emissions from a formation. This is an indicator of the radioactive material content of the rock. The log is most frequently used as an indication of the shale or clay content of a formation.

gas drilling *n*: see *air drilling.*

gas seep *n*: see *seep.*

geologist *n*: a scientist who gathers and interprets data pertaining to the rocks of the Earth's crust.

geology *n*: the science of the physical history of the Earth and its life, especially as recorded in the rocks of the crust.

geophone *n*: an instrument placed on the surface that detects vibrations passing through the Earth's crust. It is used in conjunction with seismography. See *hydrophone.*

geophysicist *n*: one who studies the physics of the way sound energy travels through the layers of the earth. The reflections of sound energy from the layers of rock in the earth can be interpreted as the buried structures that might trap oil and gas.

Geronimo *n*: see *safety slide.*

gooseneck *n*: the curved connection between the rotary hose and the swivel. See *swivel.*

grapple *n*: a device that holds by friction any object placed inside so long as the grapple is the correct diameter. The grapple normally is a flexible spiral metal band that slides over an object and then constricts around the object when the tool is pulled upwards holding the object by friction.

gravel pack *n*: a screen placed in the wellbore and the surrounding annulus that is packed with gravel of a specific size designed to prevent the passage of formation sand.

guide shoe *n*: a short, heavy, cylindrical section of steel filled with concrete and rounded at the bottom, which is placed at the end of the casing string. A passage through the center of the shoe allows drilling fluid to pass up into the casing while it is being lowered and allows cement to pass out during cementing operations. The rounded profile or nose helps guide the casing past ledges or around bends in the well. Also called casing shoe.

gusher *n*: (obsolete) an oilwell that has come in with such great pressure that the oil jets out of the well like a geyser. A gusher is actually a blowout and is extremely wasteful of reservoir fluids and drive energy. See *blowout*.

H

HCL *abbr*: hydrochloric acid.

heave compensator *n*: a device much like a very large shock absorber that is suspended in the derrick above the drill string. The piston movement of the device compensates for the vessel motion allowing the suspended drill string to remain unaffected by the motion of the vessel.

heavy lift vessel *n*: a very large ship that can be partially submerged in order to float another vessel or load on the transport deck. Once in position the load is picked up above the water when the ship rises. These special ships are used to transport semisubmersible drilling rigs and other large loads from one location to another.

hexagonal kelly *n*: a hexagonal-shaped kelly. See *kelly*.

HFL *abbr*: hydrofluoric acid.

hoist *n*: 1. an arrangement of pulleys and wire rope or chain used for lifting heavy objects; a winch or similar device. 2. the drawworks. *v*: to raise or lift.

hole *n*: in drilling operations, the wellbore or borehole. See *borehole, wellbore*.

hook *n*: a large, hook-shaped device from which the swivel is suspended. It is designed to carry maximum loads ranging from 100 to 650 tons (90 to 590 tonnes) and turns on bearings in its supporting housing.

hook load *n*: the weight of the drill stem that is suspended from the hook.

horizontal *n*: deviation of 90° from vertical.

horizontal well *n*: a wellbore that has been deviated to essentially a horizontal attitude of 90° from vertical.

horsepower *n*: a unit of measure of work done by a machine. One horsepower equals 33,000 foot-pounds per minute. (Kilowatts are used to measure power in the international, or SI, system of measurement.)

hot wire *n*: a heated filament that is used to combust small amounts of natural gas recovered from the drilling mud. The resulting heat can be calibrated as gas units which are a direct indication of natural gas in the well.

hull *n*: the outer shell of a floating vessel.

hydraulic *adj*: 1. of or relating to water or other liquid in motion. 2. operated, moved, or affected by water or liquid.

hydraulic fracturing *n*: see *formation fracturing*.

hydraulic jars *n pl*: a drilling tool that can be placed in the drill string and used to deliver a blow or shock to the drill string. The tool has a hydraulic piston inside that can be compressed or cocked by the correct movement of the drill string. The cocked piston can be released driving an internal hammer against an anvil. The impact creates a very strong force that acts along the drill string that can be directed either upward or downward. This force can help free a stuck drill string.

hydrocarbons *n pl*: organic compounds of hydrogen and carbon whose densities, boiling points, and freezing points increase as their molecular weights increase. Although composed of only two elements, hydrocarbons exist in a variety of compounds because of the strong affinity of carbon atoms for other atoms and for itself. Petroleum is a mixture of many different hydrocarbons.

hydrochloric acid (HCL) *n*: an aqueous solution of hydrogen chloride; a strongly corrosive acid used mainly for dissolving limestone.

hydrocyclone *n*: a cone that develops centrifugal force when fluids are pumped through it. The centrifugal force is used to separate solids like sand and silt from the drilling mud.

hydrofluoric acid (HFL) *n*: a colorless, very corrosive acid that dissolves clay and sandstone.

hydrogas *n*: another term for liquefied petroleum gas (LPG).

hydrogen sulfide *n*: a compound of two hydrogen atoms bonded with one sulfur atom (H_2S). This is a gas that can be a contaminant in oil and natural gas. H_2S is a poisonous gas and can be lethal at concentrations above 150 ppm. At very low concentrations, the gas has a rotten egg smell. At higher concentrations, the gas is odorless because the gas has killed the sense of smell.

hydrophone *n*: a device trailed in an array behind a boat in offshore seismic exploration that is used to detect sound reflections, convert them to electric current, and send them through a cable to recording equipment on the boat. See *geophone*.

I

IADC *abbr*: International Association of Drilling Contractors.

impermeable *adj*: not allowing the passage of fluid. See *permeability*. Compare *permeable*.

inclination *n*: the angle of the borehole from vertical. Measured in degrees with vertical being 0° and horizontal being 90°.

independent *n*: a nonintegrated oil company or an individual whose operations are in the field of petroleum production, excluding transportation, refining, and marketing.

infill drilling *n*: drilling wells between known producing wells to exploit the resources of a field to best advantage.

infill well *n*: a well drilled between known producing wells to exploit the reservoir better.

inland barge *n*: a floating offshore drilling structure consisting of a barge on which the drilling equipment is constructed which can be moved to different locations. When stationed on the drill site, the barge can be anchored in the floating mode or submerged to rest on the bottom. Typically, inland barge rigs are used to drill wells in marshes, shallow inland bays, and areas where the water is not too deep. Also called swamp barge, inland barge rig. See *floating offshore drilling rig.*

insert *n*: a cylindrical object, rounded, blunt, or chisel-shaped on one end and usually made of tungsten carbide, that is inserted in the cones of a bit, the cutters of a reamer, or the blades of a stabilizer to form the cutting element of the bit or the reamer or the wear surface of the stabilizer. Also called a compact.

intermediate casing *n*: the string of casing set in a well after the surface casing but before production casing is set to keep the hole from caving and to seal off troublesome formations. In deep wells, one or more intermediate strings may be required. Also called protection casing.

intermediate string *n*: see *intermediate casing.*

International Association of Drilling Contractors (IADC) *n*: an organization of drilling contractors that sponsors or conducts research on education, accident prevention, drilling technology, and other matters of interest to drilling contractors and their employees. www.iadc.org

Iron Roughneck™ *n*: a manufacturer's name for a floor-mounted combination of a spinning wrench and a torque wrench. The Iron Roughneck™ moves into position hydraulically and eliminates the manual handling involved with making up or breaking out pipe.

J

jackup *n*: a mobile bottom-supported offshore drilling structure with columnar or open-truss legs that support the deck and hull. When positioned over the drilling site, the bottoms of the legs rest on the seafloor. A jackup rig is towed or propelled to a location with its legs up. Once the legs are firmly positioned on the bottom, the deck and the hull height are adjusted and leveled. Also called self-elevating drilling unit.

jet *n*: 1. a hydraulic device operated by a centrifugal pump used to clean the mud pits, or tanks, and to mix mud components. 2. in a perforating gun using shaped charges, a highly penetrating, fast-moving stream of exploded particles that forms a hole in the casing, cement, and formation. 3. nozzle in a bit.

jetted *v*: the act of pumping waste fluids from one pit to another. A common practice where shale and reserve pits are used.

joint *n*: a single length (from 16 feet to 45 feet, or 5 metres to 14.5 metres, depending on its range length) of drill pipe, drill collar, casing, or tubing that has threaded connections at both ends.

joint of pipe *n*: a length of drill pipe or casing. Both come in various lengths. See *range length.*

junk *n*: metal debris lost in a hole.

junk mills *n pl*: a drag bit that has tungsten carbide as the cutting structure. The very hard tungsten carbide can drill away or mill pieces of metal that have been left in a well as junk.

K

kelly *n*: the heavy steel tubular device, four- or six-sided, suspended from the swivel through the rotary table and connected to the top joint of drill pipe to turn the drill stem as the rotary table turns. It has a bored passageway that permits fluid to be circulated into the drill stem and up the annulus.

kelly bushing *n*: see *kelly drive bushing.*

kelly drive bushing *n*: a device that fits into the master bushing of the rotary table and through which the kelly runs. When the master bushing rotates the kelly drive bushing, the kelly drive bushing rotates the kelly and the drill stem attached to the kelly.

kelly joint *n*: see *kelly.*

kelly spinner *n*: a pneumatically operated device mounted on top of the kelly that, when actuated, causes the kelly to turn, or spin. It is used when making up or breaking out the kelly from the drill string.

kerogen *n*: the organic source of hydrocarbons.

kick *n*: an entry of water, gas, oil, or combination of those fluids into the wellbore during drilling. It occurs because the pressure exerted by the column of drilling fluid is not great enough to overcome the pressure exerted by the fluids in the formation drilled.

kick off *v*: to deviate a wellbore from the vertical, as in directional drilling.

kilometres *n*: a unit of length or distance in the metric and SI units systems. One kilometre is equal to 1,000 metres or 3,280.84 feet.

L

lag *n*: the time delay that it takes the cuttings and gas samples to reach the surface from the bottom of the well.

land rig *n*: any drilling rig that is located on dry land. Compare *offshore rig*.

latch on *v*: to attach elevators to a section of pipe to pull it out of or run it into the hole.

latch tongs *n*: see *tongs*.

lead tongs *n pl*: pipe tongs suspended in the derrick or mast and operated by a chain or a wire rope connected to the makeup cathead or the breakout cathead. See *makeup tongs*.

lease *n*: an agreement that gives the right to explore land for oil, gas, and sometimes other minerals and to extract them from the ground.

limestone *n*: an evaporite mineral (rock) type made up of calcium carbonate ($CaCO_3$). The mineral has a specific gravity of 2.71 and is soluble in hydrochloric acid.

liner *n*: a string of pipe used to case open hole below existing casing. A liner extends from the setting depth up into another string of casing, usually overlapping about 100 feet (30 metres) into the upper string.

liquefied natural gas (LNG) *n*: a natural gas that has been cooled to –260°F at atmospheric pressure. At this very cold temperature, the gas becomes a liquid. If allowed to return to a normal temperature the liquid vaporizes. LNG occupies about 1/600th of the volume it occupied as a gas and the liquid can be shipped in insulated containers.

liquefied petroleum gas (LPG) *n*: a propane or butane gas that has been pressurized to the point where it turns into a liquid. LPG must be stored and shipped in pressurized containers. If the pressure is released, the liquid propane or butane vaporizes. See *hydrogas*.

liquid *n*: a state of matter in which the shape of the given mass depends on the containing vessel, but the volume of the mass is independent of the vessel. A liquid is a fluid that is almost incompressible.

LNG *abbr*: *liquefied natural gas*.

location *n*: the place where a well is drilled. Also called the well site.

log *n*: a systematic recording of data, such as a driller's log, a mud log, an electrical well log, or a nuclear log. *v*: to record data.

logging while drilling (LWD) *n*: logging measurements obtained by measurement-while-drilling techniques as the well is being drilled.

LPG *abbr*: *liquefied petroleum gas*.

LWD *abbr*: *logging while drilling*.

M

m *abbr*: *metre*.

made up *v*: see *make up*.

magnetometer *n*: a device that measures the strength of the Earth's magnetic field. Used in directional drilling to determine the compass bearing of the bottom of the well.

major *n*: a large oil company, such as ExxonMobil or Chevron, that not only produces oil, but also transports, refines, and markets it and its products.

make a connection *v*: to attach a joint of drill pipe onto the drill stem suspended in the wellbore to permit deepening the wellbore by the length of the joint (usually about 30 feet, or 9 metres.)

make up *v*: 1. to assemble and join parts to form a complete unit (e.g., to make up a string of casing). 2. to screw together two threaded pieces. 3. to mix or prepare (e.g., to make up a tank of mud).

makeup *adj*: added to a system (e.g., makeup water used in mixing mud).

make up a joint *v*: to screw a length of pipe into another length of pipe.

makeup cathead *n*: a device that is attached to the shaft of the drawworks and used as a power source for screwing together joints of pipe. It is the automatic cathead located on the driller's side of the drawworks. See *cathead*. Compare *breakout cathead*.

makeup tongs *n pl*: tongs used for screwing one length of pipe into another for making up a joint. See *lead tongs, tongs*.

making hole *v*: to deepen the hole made by the bit, i.e., to drill ahead.

male connection *n*: a pipe, a coupling, or a tool that has threads on the outside so that it can be joined to the threads on the inside of a female connection. Also called the pin. Compare *female connection*.

manifold *n*: an accessory system of piping to a main piping system (or another conductor) that serves to divide a flow into several parts, to combine several flows into one, or to reroute a flow to any one of several possible destinations.

marine crew *n*: the crew of officers and sailors who are responsible for the maritime activities of a floating drilling vessel.

marine riser *n*: see *marine riser system*.

marine riser connector *n*: a fitting on top of the subsea blowout preventers to which the riser pipe is connected.

marine riser system *n*: the pipe and special fittings used on floating offshore drilling rigs to establish a seal between the top of the wellbore, which is on the ocean floor, and the drilling equipment, located above the surface of the water. A riser pipe serves as a guide for the drill stem from the drilling vessel to the wellhead and as a conductor of drilling fluid from the well to the vessel.

mast *n*: a portable derrick that is capable of being raised as a unit. Compare *derrick*.

master bushing *n*: a device that fits into the rotary table to accommodate the slips and drive the kelly bushing so that the rotating motion of the rotary table can be transmitted to the kelly. Also called rotary bushing.

matrix *n*: the shape of the bit.

measurement while drilling (MWD) *n*: 1. directional or other surveying during routine drilling operations to determine the angle and direction by which the wellbore deviates from the vertical. The measured data are encoded and transmitted to the surface recorder as a series of pressure pulses in the mud. The surface recorder sends the code to a computer where it is converted back into the measured data. 2. any system of measuring and transmitting to the surface downhole information during routine drilling operations. See *logging while drilling*.

mechanical cathead *n*: see *cathead*.

mechanical rig *n*: a drilling rig in which the source of power is one or more internal-combustion engines and in which the power is distributed to rig components through mechanical devices (such as chains, sprockets, clutches, and shafts). Also called a power rig. Compare *electric rig*.

methane *n*: the simplest hydrocarbon compound, methane is an odorless and colorless gas made up of four hydrogen atoms bonded to a single carbon atom (CH_4). Methane is the largest single component by volume in natural gas.

metre (m) *n*: the fundamental unit of length in the inter-national system of measurement (SI). It is equal to about 3.28 feet, 39.37 inches, or 100 centimetres.

metreage contract *n*: see *footage rate*.

metric system *n*: a decimal system of weights and measures based on the metre as the unit of length, the gram as the unit of weight, the cubic metre as the unit of volume, the litre as the unit of capacity, and the square metre as the unit of area. The international system of measurement (SI) is based on the metric system.

metric ton *n*: a measurement of mass equal to 1,000 kilograms or 2,204.6 pounds. In some oil-producing countries, production is reported in metric tons. One metric ton is equivalent to about 7.4 barrels (42 U.S. gallons = 1 barrel) of crude oil, but this depends on the density of the oil. In the SI system it is called a tonne.

mill *n*: a downhole drag bit with rough, sharp, extremely hard tungsten carbide cutting surfaces for removing metal by grinding or cutting. They are also called junk mills or reaming mills depending on use. *v*: to use a mill to cut or grind metal objects that must be removed from a well.

mineral rights *n pl*: the rights of ownership, conveyed by deed, of gas, oil, and other minerals beneath the surface of the earth. In the United States, mineral rights are the property of the surface owner unless disposed of separately. In most other countries the mineral rights are the property of the State.

mobile offshore drilling unit (MODU) *n*: a drilling rig used to drill offshore exploration and development wells which floats on the surface of the water when being moved from one drill site to another. When drilling, it may float or be bottom supported.

MODU *abbr*: *mobile offshore drilling unit*.

monkeyboard *n*: the derrickhand's working platform at the correct height in the derrick for handling the top of the pipe. As pipe is run into or out of the hole, the derrickman must handle the top end of the pipe (90 feet (27 metres) or higher in the derrick or mast).

mousehole *n*: an opening through the rig floor, usually lined with pipe, into which a length of drill pipe is placed temporarily for later connection to the drill string.

mud *n*: the liquid circulated through the wellbore during rotary drilling operations. In addition to its function of bringing cuttings to the surface, drilling mud cools and lubricates the bit and the drill stem, protects against blowouts by holding back subsurface pressures, and deposits a mud cake on the wall of the borehole to prevent loss of fluids to the formation. See *drilling fluid*.

mud centrifuge *n*: a device that uses centrifugal force to separate small solid components from liquid drilling fluid.

mud cleaner *n*: a cone-shaped desander, a hydrocyclone, designed to remove very fine solid particles from oil-based drilling mud without removing the barite weighting material

mud engineer *n*: an employee of a drilling fluid supply company whose duty it is to test and maintain the drilling mud properties that are specified by the operator.

mud hopper *n*: a large funnel- or cone-shaped device into which dry components (such as powdered clay or cement) can be poured to mix uniformly with water or other liquids.

mud line *n*: 1. in offshore operations, the seafloor. 2. a mud return line.

mud logger *n*: an employee of a mud logging company who performs mud logging.

mud logging *n*: the recording of information derived from examination and analysis of formation cuttings made by the bit and of mud circulated out of the hole.

mud logging company *n*: a service company that monitors and logs the content of the drilling mud as it returns from the well.

mud motor *n*: see *downhole motor*.

mud pit *n*: originally, an open pit dug in the ground to hold drilling fluid or waste materials discarded after the treatment of drilling mud. For some drilling operations, mud pits are used for suction to the mud pumps, settling of drilled cuttings, and storage of reserve mud. Steel tanks are commonly used now but still referred to as pits. Offshore, the term "mud tanks" is preferred.

mud pump *n*: a large, high-pressure positive displacement pump used to circulate the mud on a drilling rig. Also called a slush pump.

mud return line *n*: a trough or pipe that is placed between the surface connections at the wellbore and the shale shaker and through which drilling mud flows on its return to the surface from the hole. Also called a flow line.

mud tank *n*: one of a series of open tanks, usually made of steel plate, through which the drilling mud is cycled for treatment and to remove sand and fine sediments. Also called mud pit.

mud weight *n*: a measure of the density of a drilling fluid expressed as pounds per gallon, pounds per cubic foot, or kilograms per cubic metre. Mud weight is directly related to the amount of pressure the column of drilling mud exerts at the bottom of the hole.

mule shoe *n*: a cut lip on the overshot that acts as a guide and allows the overshot to be rotated over the fish.

MWD *abbr*: *measurement while drilling*.

N

natural gas *n*: a highly compressible, highly expandable mixture of hydrocarbons with a low specific gravity and occurring naturally in a gaseous form. Besides hydrocarbon gases, natural gas may contain appreciable quantities of nitrogen, helium, carbon dioxide, hydrogen sulfide, and water vapor.

natural gas liquid (NGL) *n*: liquid hydrocarbons that condense from natural gas when the pressure and temperature of the gas are reduced.

neutron log *n*: a tool that can be used as an indirect indication of formation porosity. The log records the radioactive response from a formation to the bombardment of the formation with neutrons. The neutrons are slowed by collisions with large atoms like hydrogen. Hydrogen occurs in water, oil, and gas that occur only in the pore space of the rock. The more hydrogen present the faster the neutrons are slowed down and the larger the porosity is inferred to be. Unlike the density log, which also is a porosity indicator, the neutron log can be used in open holes and in wells that have been cased.

NGL *abbr*: *natural gas liquid*.

night toolpusher *n*: an assistant whose duty hours are typically during nighttime hours on a mobile offshore drilling unit.

nippled up *v*: 1. to assemble the blowout preventer stack or other wellhead components on the wellhead at the surface. 2. to assemble.

nozzle *n*: a passageway through jet bits that causes the drilling fluid to be ejected from the bit at high velocity. The jets of mud clear the bottom of the hole and help keep the cutting structures of the bit clean.

O

off-bottom weight *n*: the weight of the drill string that is suspended by the derrick before the bit is allowed to touch the bottom of the well.

offshore *n*: that geographic area that lies seaward of the coastline. In general, the term coastline means the line of ordinary low water along that portion of the coast that is in direct contact with the open sea or the line marking the seaward limit of inland waters.

offshore drilling *n*: drilling for oil or gas in an ocean, gulf, or sea. A drilling unit for offshore operations may be a mobile floating vessel with a ship or barge hull, a semisubmersible or submersible base, a self-propelled or towed structure with jacking legs (jackup drilling rig), or a permanent structure used as a production platform when drilling is completed.

offshore drilling rig *n*: see *offshore rig*.

offshore installation manager (OIM) *n*: a qualified and certified person with marine and drilling knowledge who is in charge of all operations on a MODU.

offshore production platform *n*: an immobile offshore structure from which wells are produced.

offshore rig *n*: any of various types of drilling structures designed for use in drilling wells in oceans, seas, bays, gulfs, and so forth. Offshore rigs include platforms, jackup drilling rigs, semisubmersible drilling rigs, and drillships. Compare *land rig*.

oil *n*: a simple or complex liquid mixture of hydrocarbons that can be refined to yield gasoline, kerosene, diesel fuel, and various other products.

oil-base mud *n*: a drilling fluid in which oil is the continuous phase and which contains from less than 2 percent and up to 10 percent water. The oil can be diesel, mineral, or synthetic oil.

oilfield *n*: referring to an area where oil is found. May also include the oil reservoir, the surface and wells, and production equipment.

oil seep *n*: a surface location where oil appears, the oil having permeated its subsurface boundaries and accumulated in small pools or rivulets. Also called oil spring.

oilwell *n*: a well from which oil is obtained.

OIM *abbr*: *offshore installation manager*.

on-bottom weight *n*: the weight of the drill string suspended by the derrick after the bit has been allowed to touch the bottom of the well. The difference between the off-bottom and on-bottom weight is the weight that has been placed on the bit to make it drill.

open *adj*: 1. a wellbore with no casing. 2. a hole with no drill pipe or tubing suspended in it.

open hole *n*: any wellbore in which casing has not been set.

open-hole completion *n*: to leave the producing formation open by setting the production casing at the top of the reservoir.

open-hole fishing *n*: the procedure of recovering lost or stuck equipment in an uncased wellbore.

operating company *n*: see *operator*.

operator *n*: the person or company actually operating an oilwell that hires a drilling contractor. The term operator carries the connotation of authority over the well from beginning to end. The operator oversees all operations during the life of the well.

outpost well *n*: a well located outside the established limits of a reservoir, i.e., a step-out well.

overshot *n*: a fishing tool attached to tubing or drill pipe and lowered over the outside wall of pipe lost or stuck in the wellbore. A friction device in the overshot, usually a spiral grapple, firmly grips the pipe, allowing the fish to be pulled from the hole.

P

P&A *abbr*: plug and abandon.

packer *n*: a piece of downhole equipment that consists of a sealing device, a holding or setting device, and an inside passage for fluids. It is used to block the flow of fluids through the annular space between pipe and the wall of the wellbore by sealing off the space between them. A packing element expands to prevent fluid flow except through the bore of the packer and tubing.

PDC *abbr*: polycrystalline diamond compact.

PDC bit *n*: a special type of man-made diamond drag bit. Polycrystalline diamond inserts, or compacts, are embedded into a matrix on the bit.

penetration rate *n*: see *rate of penetration*.

perforate *v*: to pierce the casing wall and cement of a wellbore to provide holes through which formation fluids may enter or to provide holes in the casing or tubing so that materials may be introduced into the annulus.

perforated liner *n*: a liner that either has had holes predrilled or cut or that are shot in it by a perforating gun. See *liner*.

perforating gun *n*: a device fitted with shaped charges or bullets that is lowered to the desired depth in a well and fired to create penetrating holes in casing, cement, and formation.

perforation *n*: a hole made in the casing, cement, and formation through which formation fluids enter a wellbore. Usually, several perforations are made at a time.

permeability *n*: 1. a measure of the ease with which a fluid flows through the connecting pore spaces of a rock. The unit of measurement is the millidarcy (square metres). 2. fluid conductivity of a porous medium. 3. ability of a fluid to flow within the interconnected pore network of a porous medium.

permeable *adj*: allowing the passage of fluid. See *permeability*. Compare *impermeable*.

personal protective equipment (PPE) *n*: items issued to individuals at the rig site and used for the individual's personal protection. The items are safety glasses, hearing protection, a hard hat, and steel toed boots or shoes. Other items may be special coveralls, face shields, gloves, and safety harnesses.

petrochemical feedstock *n*: hydrocarbons used as the raw material for petrochemicals such as pharmaceuticals, plastics, synthetic rubber, and the like.

petroleum *n*: a substance occurring naturally in the earth in solid (tar), liquid (crude oil), or gaseous (natural gas) state and composed mainly of mixtures of chemical compounds of carbon and hydrogen, with or without other nonmetallic elements such as sulfur, oxygen, and nitrogen. In some cases, especially in the measurement of oil and gas, petroleum refers only to oil—a liquid hydrocarbon—and does not include natural gas or gas liquids such as propane and butane. The API prefers that petroleum mean crude oil and not natural gas or gas liquids.

petroleum geology *n*: the study of oil- and gas-bearing rock formations. It deals with the origin, the occurrence, the movement, and the accumulation of hydrocarbons.

pick up *v*: 1. to use the drawworks to lift the bit (or other tool) off bottom by raising the drill stem. 2. to use an air hoist or friction cathead to lift a tool, a joint of drill pipe, or other piece of equipment.

piercement salt dome *n*: a salt dome pushed up so that it penetrates the overlying sediments, leaving them truncated.

pile driver *n*: a ram that is used like a large hammer to drive posts, piles, and conductor casing into the earth. The ram is attached to a hydraulic cylinder and driven by a power source like a diesel engine.

pin *n*: 1. the male section of a tool joint. 2. on a bit, the bit shank that screws into a bit sub or drill collar.

pipe *n*: a long, hollow cylinder, usually steel, through which fluids are conducted.

pipe rack *n*: a horizontal support for tubular goods.

pipe racker *n*: a pneumatic or hydraulic device that, on command from an operator, either picks up pipe from a rack or from the side of the derrick and lifts it into the derrick or takes pipe from out of the derrick and places it on the rack or places it to the side of the derrick. This machine takes the place of the derrickman and in some cases the floormen as well.

pipe ram *n*: a sealing component for a blowout preventer that closes the annular space between the pipe and the blowout preventer or wellhead.

pipe ram preventer *n*: a blowout preventer that uses pipe rams as the closing elements. See *pipe ram*.

pipe tally *n*: a written or digital record of the individual dimensions (inner and outer diameter and length) of a string of pipe that has been placed in a well.

pipe tally book *n*: a book where the pipe tally is kept.

pipe tongs *n pl*: see *tongs*.

pit level *n*: height of drilling mud in the mud tanks, or pits.

plastic *n*: 1. materials made from hydrocarbons in petrochemical plants. 2. a characteristic of certain materials where the shape of the material can be changed permanently by applying the correct stress to the material.

platform *n*: an immobile offshore structure from which development wells are drilled and produced. Platforms may be built of steel or concrete and may be rigid or compliant. Rigid platforms, which rest on the seafloor, are the concrete gravity platform and the steel-jacket platform. Platforms, which are used in deeper waters and yield to water and wind movements, are the guyed-tower platform and the tension-leg platform.

platform jacket *n*: a support that is firmly secured to the ocean floor and to which the legs of a platform are anchored.

platform rig *n*: a drilling rig placed on an offshore platform.

play *n*: 1. the extent of a petroleum-bearing formation. 2. the activities associated with petroleum development in an area.

plug and abandon (P&A) *v*: to place cement and/or mechanical plugs into a dry hole or a depleted well to seal and abandon it.

plug container *n*: see *cementing head*.

pneumatic *adj*: operated by air pressure.

polycrystalline diamond compact (PDC) *n*: a disk (a compact) of very small synthetic diamonds, metal powder, and tungsten carbide powder that are used as cutters on PDC bits. Compare *thermally stable polycrystalline diamond bit*.

pontoons *n pl*: floats or outriggers placed on a semisubmersible rig that can be partially flooded with seawater to control the draft of the vessel.

pore *n*: an opening or space within a rock or mass of rocks, usually small and often filled with some fluid (water, oil, gas, or all three). Compare *vug*.

porosity *n*: 1. the condition of being porous (such as a rock formation). 2. the ratio of the volume of empty space to the total (bulk) volume of rock in a formation, indicating how much fluid a rock can hold.

porous *adj*: having pores, or tiny openings, as in rock.

posted barge *n*: a submersible rig for shallow water or marsh locations. The rig has the drilling platform raised above a barge hull by steel posts. The rig is towed to location and the barge hull flooded and sunk to the bottom. The steel posts keep the rig floor above water.

power rig *n*: see *mechanical rig*.

power swivel *n*: see *top drive*.

PPE *abbr: personal protective equipment.*

pressure *n*: the force that a fluid (liquid or gas) exerts uniformly in all directions within a vessel, a pipe, a hole in the ground, and so forth, such as that exerted against the inner wall of a tank or that exerted on the bottom of the wellbore by a fluid. Pressure is expressed in terms of force exerted per unit of area, as pounds per square inch, or in kilopascals.

preventer *n*: shortened form of blowout preventer. See *blowout preventer.*

prime mover *n*: an internal-combustion engine or a motor that is the source of power for driving a machine or machines.

production *n*: 1. the part of the petroleum industry that deals with bringing the well fluids to the surface and separating them and with storing, gauging, and otherwise preparing the product for the pipeline. 2. the amount of oil or gas produced in a given period.

production casing *n*: the last string of casing set in a well, inside of which is usually suspended a tubing string.

production tubing *n*: tubing used in the completion of a well that fits inside the production casing.

production wellhead *n*: the control valves through which oil and/or natural gas are produced. See *wellhead.*

proppant *n*: a material used to hold a fracture open when the pressure is removed. The material can be sand or other material. See *propping agent.*

propping agent *n*: a granular substance (sand grains, bauxite, or other material) that is carried in suspension by the fracturing fluid and that serves to keep the cracks open when fracturing pressure is released after a fracture treatment.

pulley *n*: a wheel with a grooved rim, used for pulling or hoisting. Also called a sheave.

pull line *n*: a length of wire rope one end of which is connected to the end of the tongs and the other end of which is connected to the chain on the automatic cathead on the drawworks. When the driller actuates the cathead, it takes in the tong line and exerts force on the tong to either make up or break out drill pipe.

pull singles *v*: to remove the drill stem from the hole by disconnecting each individual joint.

pump *n*: a device that increases the pressure on a liquid.

Q

quadruple *n*: the modern term for a stand of four joints of the drill string left connected together. Also called fourble (obsolete).

R

rack *n*: framework for supporting or containing a number of loose objects, such as pipe. *v*: 1. to place on a rack. 2. to use as a rack.

racked back *v*: the process of taking a stand of drill pipe or drill collars and placing the top inside the fingers of the fingerboard for temporary storage. The stands are placed one after another or racked back in the fingerboard.

ram *n*: the closing and sealing component on a blowout preventer. One of three types—blind, pipe, or shear—may be installed in several preventers mounted in a stack on top of the wellbore.

ram blowout preventer *n*: a blowout preventer that uses rams to seal off pressure on a hole that is with or without pipe. Also called ram preventer. Compare *annular blowout preventer.*

ram preventer *n*: see *ram blowout preventer.*

range of length *n*: a grouping of pipe lengths. See *joint.*

rate of penetration (ROP) *n*: a measure of the speed at which the bit drills into formations, usually expressed in feet (metres) per hour or minutes per foot (metre).

rathole *n*: a hole in the rig floor, which is lined with a pipe that projects above the floor and into which the kelly and swivel are placed when hoisting operations are in progress.

rathole rig *n*: a small, usually truck-mounted, rig used to drill the rathole and the mousehole for the main drilling rig which will be moved in later. A rathole rig may also drill the top part of the hole, the conductor hole, before the main rig arrives on location.

reel *n*: a revolving device (such as a flanged cylinder) for winding or unwinding something flexible (such as rope or wire).

remotely operated vehicle (ROV) *n*: a small unmanned submarine generally attached to the floating drilling rig by cables. The small submarine has a camera for viewing the underwater operation and has an arm with a gripper or claw that can manipulate some subsea controls on wellheads or other underwater equipment. The operator has a command station on board the rig where remote control of the submarine is achieved.

repeat formation tester (RFT) *n*: a wireline conveyed formation evaluation tool that can obtain formation pressures and fluid samples from open-hole sections of the well.

reserve pit *n*: a waste pit, an excavated earthen-walled pit. It may be lined with plastic or other material to prevent soil contamination.

reservoir *n*: a subsurface, porous, permeable rock body in which oil or gas has accumulated.

reservoir pressure *n*: the pressure within the reservoir at any given time.

reservoir rock *n*: a permeable rock that may contain oil or gas in appreciable quantity and through which fluids can migrate.

rig *n*: the derrick or mast, drawworks, and attendant surface equipment of a drilling unit.

rig crewmember *n*: see *floorhand*.

rig floor *n*: the area immediately around the rotary table and extending to each corner of the derrick or mast. Also called derrick floor, drill floor.

rigged up *v*: see *rigging up*.

rigging down *v*: 1. to dismantle a drilling rig and auxiliary equipment following the completion of drilling operations. Also called tear down. 2. in general, to disassemble.

rigging up *v*: 1. to prepare the drilling rig for making hole, i.e., to install tools and machinery before drilling is started. 2. in general, to prepare for an operation.

rig hand *n*: a crewmember who is part of a drilling crew. See *derrickman, driller, floorhand*.

rig manager *n*: an employee of a drilling contractor who is in charge of the entire drilling crew and the drilling rig, providing logistics support to the rig crew and liaison with the operating company. See *toolpusher*.

rig superintendent *n*: see *toolpusher*.

rig supervisor *n*: see *toolpusher*.

riser pipe *n*: see *marine riser*.

rock *n*: a hardened aggregate of different minerals. Rocks are divided into three groups on the basis of their mode of origin: igneous, metamorphic, and sedimentary.

rock oil *n*: see *petroleum*.

roller cone *n*: a drilling bit made of three cones that are mounted on extremely rugged bearings. The surface of each cone has rows of steel teeth or rows of tungsten carbide inserts. When the bit is rotated, the cones roll across the rock surface. This action causes the teeth on the cones to first crush and then shear rock into fragments or cuttings. Also called rock bits.

roller cone bit *n*: see *roller cone*.

ROP *abbr*: rate of penetration.

rotary *n*: the machine used to impart rotational power to the kelly and drill stem while permitting vertical movement of the pipe for rotary drilling.

rotary drilling *n*: a drilling method in which a hole is drilled by a rotating bit to which a downward force is applied. The bit is fastened to and rotated by the drill stem, which also provides a passageway through which the drilling fluid is circulated.

rotary drilling rig *n*: see *rotary rig*.

rotary helper *n*: see *floorhand*.

rotary hose *n*: a reinforced flexible tube on a rotary drilling rig that conducts the drilling fluid from the standpipe to the swivel and kelly or to the top drive. Also called *kelly hose, mud hose*.

rotary rig *n*: a drilling rig that has a system that rotates a bit and, at the same time, has a system that continuously circulates drilling fluid while drilling is going on.

rotary speed *n*: the speed, measured in revolutions per minute, at which the rotary table is operated.

rotary steerable assembly *n*: a directional drilling tool that allows the path of the wellbore to be measured and the position changed while rotating the drill string.

rotary table *n*: the principal component of a rotary, or rotary machine, used to turn the drill stem and support the drilling assembly.

rotary-table system *n*: a series of devices that provide a way to rotate the drill stem and bit. Basic components consist of a turntable, master bushing, kelly drive bushing, kelly, and a swivel.

rotating components *n pl*: the parts of the drilling or workover rig that are designed to turn or rotate the drill stem and bit—swivel, kelly, kelly bushing, master bushing, and rotary table.

rotating head *n*: a device that is placed on top of the drilling wellhead that seals the wellbore while still allowing the drill string to be rotated. A rotating head resembles an annular blowout preventer except that the sealing element has a bearing that allows it to rotate.

rotor *n*: the rotating inner shaft of a motor. The inner shaft rotates inside a fixed housing called the stator. The rotor is caused to turn inside the stator by some applied force. In the case of an electric motor, the force is created by the electric current flowing though the stator. In the case of a drilling motor, it is created by the pressure from the drilling mud that is pumped through the motor.

roughneck *n*: see *floorhand*.

round trip *n*: the action of pulling out and subsequently running back into the hole a string of drill pipe or tubing. Also called tripping.

roustabout *n*: a worker on an offshore rig who handles the equipment and supplies that are sent to the rig from the shore base. The head roustabout is very often the crane operator.

ROV *abbr*: remotely operated vehicle.

royalty *n*: a share of the money made from the sale of oil or gas that is paid to the mineral owner.

run casing *v*: to lower a string of casing into the hole. Also called to run pipe.

run in *v*: to go into the hole with tubing, drill pipe, and so forth.

running quicksand *n*: extremely unconsolidated sand that will not hold its shape when drilled. The loose sand caves in quickly filling in the hole.

run pipe *v*: to lower a string of casing into the hole. Also called to run casing.

S

safety slide *n*: a wireline device normally mounted near the monkeyboard to afford the derrickman a means of quick exit to the surface in case of emergency. It is usually affixed to a wireline, one end of which is attached to the derrick or mast and the other end to the surface. To exit by the safety slide, the derrickman grasps a handle on it and rides it down to the ground. Also called a Geronimo or Tinkerbell line.

salt dome *n*: a dome that is caused by an intrusion of salt into overlying sediments. See *piercement salt dome*.

samples *n pl*: 1. the well cuttings obtained at designated footage intervals during drilling. From an examination of these cuttings, the geologist determines the type of rock and formations being drilled and estimates oil and gas content. 2. small quantities of well fluids obtained for analysis.

sand *n*: 1. an abrasive material composed of small quartz grains formed from the disintegration of preexisting rocks. Sand consists of particles less than 0.078 inch (2 millimetres) and greater than 0.062 inch ($^{1}/_{16}$ millimetre) in diameter. 2. sandstone.

sand reel *n*: a winch. A drum, operated by a wheel, for raising or lowering the sand pump or bailer during drilling.

sandstone *n*: a sedimentary rock composed of individual mineral grains of rock fragments between 0.002 and 0.079 inches (0.06 and 2 millimetres) in diameter and cemented together by silica, calcite, iron oxide, and so forth. Sandstone is commonly porous and permeable and a likely rock in which to find a petroleum reservoir.

scratcher *n*: a device that is fastened to the outside of casing to remove mud cake from the wall of a hole to condition the hole for cementing. By rotating or moving the casing string up and down as it is being run into the hole, the scratcher, formed of stiff wire, removes the cake so that the cement can bond solidly to the formation.

SCR *abbr*: silicon-controlled-rectification.

SCR house *n*: the control house where AC current is converted to DC current and transmitted to the various motors on the rig. SCR stands for the AC to DC conversion process known as silicon controlled rectification.

seafloor *n*: the bottom of the ocean; the seabed.

seat *n*: the point in the wellbore at which the bottom of the casing is set.

sediment *n*: in geology, buried layers of sedimentary rocks.

seep *n*: the surface appearance of oil or gas that results naturally when a reservoir rock becomes exposed to the surface, thus allowing oil or gas to flow out of fissures in the rock.

seismic *adj*: of or relating to an earthquake or earth vibration, including those artificially induced.

seismic survey *n*: an exploration method in which strong low-frequency sound waves are generated on the surface or in the water and reflect off of subsurface rock structures that may contain hydrocarbons. Interpretation of the record can reveal possible hydrocarbon-bearing formations.

seismology *n*: the study of sound transmission in layers of the earth. See *geophysicist*.

self-elevating substructure *n*: a base on which the floor and mast of a drilling rig rests and which, after it is placed in the desired location, is raised into position as a single unit, normally using hydraulic pistons.

self-propelled *adj*: having its own means of propulsion.

semisubmerged *n*: a state in which a specially designed floating drilling rig (a semisubmersible) floats just below the water's surface.

semisubmersible *n*: a floating offshore drilling unit that has pontoons and columns that, when flooded, cause the unit to submerge to a predetermined depth. Semisubmersibles are more stable than drillships and are used extensively to drill wildcat wells in rough waters such as the North Sea. See *floating offshore drilling rig*.

set back *v*: to place stands of drill pipe and drill collars in a vertical position to one side of the rotary table in the fingerboard of the derrick or mast of a drilling or workover rig. See *racked back*.

shaker *n*: shortened form of shale shaker. See *shale shaker*.

shale *n*: a fine-grained sedimentary rock composed mostly of consolidated clay or mud. Shale is the most frequently occurring sedimentary rock.

shale diapir *n*: see *diapir*.

shale pit *n*: a small earthen pit used to catch rock cuttings that are separated from the active mud volume.

shale shaker *n*: a vibrating screen used to remove cuttings from the circulating fluid in rotary drilling operations. The size of the openings in the screen should be carefully selected to be the smallest size possible to allow 100 percent flow of the fluid through the screen. Also called a shaker.

shaped charge *n*: a relatively small conical shaped container of high explosive that is loaded into a perforating gun. On detonation, the charge releases a small, high-velocity stream of particles (a jet) that penetrates the casing, cement, and formation. See *perforating gun*.

shear ram *n*: the component in a blowout preventer that cuts, or shears, through drill pipe and forms a seal against well pressure.

shear ram preventer *n*: a blowout preventer that uses shear rams as closing elements.

sheave *n*: a grooved pulley.

shut in *v*: 1. to close the valves on a well so that it stops producing. 2. to close in a well in which a kick has occurred.

sidewall core *n*: a small cylindrical rock sample (1" x 3") obtained using an electric wireline conveyed sidewall core gun. The sample is extracted by driving a core barrel into the side of the borehole wall and then pulling the barrel with the sample out.

sidewall core gun *n*: a device run into the well on an electric wireline. The gun has twenty-five empty core barrels loaded in front of individual explosive charges. The barrels are connected to the gun with steel cables. Once positioned at the correct depth, individual core barrels are fired into the side of the hole and a sample retrieved when the gun is moved. Sidewall core guns can be run in tandem to recover fifty samples or more per run into the well.

silicon-controlled-rectification (SCR) *n*: an electric process where AC current is converted to the correct voltage of DC current to power the equipment on the rig. See *electric rig*.

single *n*: a joint of drill pipe. Compare *double, quadruple, triple*.

sinker bar *n*: a heavy weight or bar placed on or near a lightweight wireline tool. The bar provides weight so that the tool will lower properly into the well.

skid *v*: to move an object by dragging it.

skid the rig *v*: to drag parts of a rig to a new location.

SL *abbr*: surface location.

slacks off *v*: to diminish hold or tension on the drill string.

slide drilling *n*: a process in which the direction of the bottom of the well is changed. The method uses a bent housing motor where the bit can be pointed in a new direction. The bit is rotated by the motor, but the drill string is not rotated. The bit drills in the new direction with the drill string sliding along behind.

slingshot substructure *n*: see *self-elevating substructure*.

slips *n pl*: wedge-shaped pieces of metal with teeth or other gripping elements that are used to prevent pipe from slipping down into the hole or to hold pipe in place.

slurry *n*: in drilling, a plastic mixture of cement and water that is pumped into a well to harden. There it supports the casing and provides a seal in the wellbore to prevent migration of underground fluids.

sonic log *n*: a wireline tool that provides an indication of formation porosity. The tool measures the time required for sound to travel from a transmitter on the tool through the formation and back to a receiver on the tool. The time required to travel this fixed distance is proportional to the volume and type of rock and the volume and type of fluid present in the formation.

sour crude *n*: oil containing hydrogen sulfide gas.

sour gas *n*: natural gas containing significant amounts of hydrogen sulfide.

spear *n*: a fishing tool used to retrieve pipe lost in a well. The spear is lowered down the hole and into the lost pipe. When weight, torque, or both are applied to the string to which the spear is attached, the slips in the spear expand and tightly grip the inside of the wall of the lost pipe.

spider *n*: a set of slips that can be engaged and disengaged by pulling a lever.

Spindletop *n*: the name of the location of the 1901 Lucas gusher near Beaumont, Texas. This famous well demonstrated that vast quantities of oil could be found and produced from a drilled well and that rotary drilling technology was superior to cable-tool technology for most drilling applications.

spinning wrench *n*: air-powered or hydraulically-powered wrench used to spin drill pipe when making up or breaking out connections.

spool *n*: the drawworks drum. Also a casinghead or drilling spool. *v*: to wind around a drum.

spud *v*: to begin drilling a well; i.e., to spud in.

spudded *v*: to begin drilling; to start the hole.

SSV *abbr*: surface safety valve

SSSV *abbr*: *subsurface safety valve*.

stab *v*: to guide the end of a pipe into a coupling or tool joint when making up a connection.

stabilizer *n*: a tool placed on a drill collar near the bit that is used, depending on where it is placed, either to maintain a particular hole angle or to change the angle by controlling the location of the contact point between the hole and the collars.

stand *n*: the connected joints of pipe racked in the derrick or mast when making a trip. On a rig, the usual stand is about 90 feet (about 27 metres) long (three lengths of drill pipe screwed together), or a triple.

standard derrick *n*: a derrick that is built piece by piece at the drilling location. Compare *mast*.

standpipe *n*: a vertical pipe rising along the side of the derrick or mast, which joins the discharge line leading from the mud pump to the rotary hose and through which mud is pumped into the hole.

stator *n*: the stationary housing inside of which a rotor turns. Used in electric and hydraulic motors. See *rotor*.

steel cone *n*: see *roller cone*.

steel-tooth bit *n*: a roller cone bit in which the surface of each cone is made up of rows of steel teeth. Also called a milled-tooth bit or milled bit.

stem *n*: see *sinker bar, swivel stem*.

step-out well *n*: a well drilled adjacent to or near a proven well to ascertain the limits of the reservoir; an outpost well.

stimulate *v*: the action of attempting to improve and enhance a well's performance by creating fractures in the rock, or using chemicals such as acid to dissolve the soluble portion of the rock or remove fluid blocks. See *acidizing, formation fracturing*.

stratigraphic trap *n*: a petroleum trap that occurs when the top of the reservoir bed is terminated by other beds or by a change of porosity or permeability within the reservoir itself. Compare *structural trap*.

structural trap *n*: a petroleum trap that is formed because of deformation (such as folding or faulting) of the reservoir rock. Compare *stratigraphic trap*.

sub *n*: a short, threaded piece of pipe used to adapt parts of the drilling string that cannot otherwise be screwed together because of differences in thread size or design.

submerged *n*: a state in which a rig that floats on the surface while being moved is in contact with the seafloor when it is in the drilling mode.

submersible *n*: a MODU that floats on the water's surface when moved from one drilling site to another. When it reaches the site, crewmembers flood compartments that submerge the lower part of the rig to the seafloor.

subsea blowout preventer *n*: a blowout preventer placed on the seafloor for use by a floating offshore drilling rig. Also called subsea BOP.

subsea engineer *n*: see *subsea equipment supervisor*.

subsea equipment supervisor *n*: an employee on a floating offshore drilling rig whose main responsibility is running, monitoring, and maintaining such subsea equipment as the blowout preventer stack, the marine riser system, and similar subsea equipment.

substructure *n*: the foundation on which the derrick or mast and usually the drawworks sit.

subsurface *adj*: below the surface of the earth (e.g., subsurface rocks).

subsurface safety valve (SSSV) *n*: a device installed in the tubing string of a producing well to shut in the flow of production if the pressure in the control line drops. The valve is placed in the tubing below either ground level or the mud line.

sulfide *n*: a chemical compound containing sulfur.

supply reel *n*: a spool that holds extra drilling line.

surface casing *n*: the first string of casing (after the conductor pipe) that is set in a well. It varies in length from a few hundred to several thousand feet (metres). Some states require a minimum length to protect freshwater sands. Compare *conductor casing*.

surface hole *n*: that part of the wellbore that is drilled below the conductor hole but above the intermediate hole. Surface casing is run and cemented in the surface hole.

surface location (SL) *n*: the physical site of the top of a well. See *bottomhole location*.

surface pipe *n*: see *surface casing*.

surface safety valve (SSV) *n*: a valve, mounted in the Christmas tree assembly, that stops the flow of fluids from the well in response to a variety of sensors.

swamp barge *n*: see *inland barge*.

swamper *n*: (slang) a helper on a truck, tractor, or other machine.

sweet crude *n*: oil containing little or no sulfur, especially little or no hydrogen sulfide.

swivel *n*: a rotary tool that is hung from the rotary hook and the traveling block to suspend the drill stem and to permit it to rotate freely. It also provides a connection for the rotary hose and a passageway for the flow of drilling fluid into the drill stem.

swivel stem *n*: a length of pipe inside the swivel that is installed to the swivel's washpipe and to which the kelly (or a kelly accessory) is attached. It conducts drilling mud from the washpipe and to the drill stem.

T

t *abbr*: tonne.

tally *v*: to measure and record the total length of pipe, casing, or tubing that is to be run in a well one joint at a time.

tap *v*: to bore a hole into an object.

tapered bowl *n*: a fitting, usually divided into two halves, that crewmembers place inside the master bushing to hold the slips.

TD *abbr*: total depth.

tectonic plate *n*: a massive, irregularly shaped slab of solid rock on the Earth's crust, generally composed of both continental and oceanic lithosphere which moves.

tensile force *n*: the rating of a tube for a tensile (axial pulling or stretching) load. The rating can be in pounds per square inch (psi). This is normally the minimum yield strength of the tube. Or, the rating can be a body yield strength in pounds where the minimum yield strength is multiplied by the wall area of the tube. For example, a J-55 grade of casing will have a minimum yield strength of 55,000 psi. If the casing has an outer diameter of 4½" and a wall thickness of 0.224 inches, the wall area is 3 square inches and the body yield strength becomes 55,000 pounds/square inch x 3 square inches = 165,000 pounds. This is the maximum force that this tube could withstand without danger of permanent damage to the tube.

tensioner *n*: a system of devices installed on a floating offshore drilling rig to maintain a constant tension on the riser pipe, despite any vertical motion made by the rig.

thermally stable polycrystalline (TSP) diamond bit *n*: a special type of drag bit that has synthetic diamond cutters that do not disintegrate at high temperatures. Compare *polycrystalline diamond compact*.

threads *n*: a profile that is machined inside one end of pipe or a tool joint and on the outside of the opposing pipe or tool joint end that allows the opposing ends to be fixed or screwed together. Much like a bolt and a nut are secured together.

thrusters *n*: equipment that keeps the floating rig dynamically positioned over the well without the use of anchors. See *dynamic positioning*.

thumper *n*: a truck used in seismic surveys that creates the energy pulse necessary by hitting or vibrating (thumping) the ground with a hydraulic ram.

Tinkerbell line *n*: see *safety slide*.

ton *n*: 1. (nautical) a volume measure equal to 35 cubic feet of seawater applied to the displacement of mobile offshore drilling rigs. 2. a measure of weight equal to 2,000 pounds (short ton). 3. (metric) a measure of weight equal to 1,000 kilograms or 2,240 pounds (long ton). Usually spelled tonne.

tongs *n pl*: the large wrenches used to make up or break out drill pipe, casing, tubing, or other pipe; variously called casing tongs, pipe tongs, and so forth, according to the specific use.

tonne (t) *n*: a mass unit in the metric system equal to 1,000 kilograms.

tool joint *n*: a heavy coupling element for drill pipe.

toolpush *n*: Canadian term for toolpusher. See *toolpusher*.

toolpusher *n*: an employee of a drilling contractor who is in charge of the entire drilling crew and the drilling rig. Also called a drilling foreman, rig manager, rig superintendent, or rig supervisor.

top drive *n*: a device similar to a power swivel that is used in place of the rotary table to turn the drill stem. Hung from the hook of the traveling block, a top drive also suspends the drill stem in the hole and includes pipe elevators. The top drive is powered by a motor or motors and includes a swivel.

top-drive system *n*: see *top drive*.

top plug *n*: a cement plug that follows cement slurry down the casing. It goes before the fluid used to displace the cement from the casing and separates the displacement fluid from the slurry. See *cementing*.

topside *n*: the decks that are placed above the water on drilling and production platforms.

torque *n*: the turning force that is applied to a lever to cause a shaft or other rotary mechanism to rotate or tend to do so. Torque is measured in units of length times force (foot-pounds, newton-metres).

total depth (TD) *n*: the maximum depth reached in a well.

tour *n*: a working shift for drilling crew or other oilfield workers. On rigs where a tour is 8 hours, they are called daylight, afternoon (or evening), and morning. Sometimes 12-hour tours are used, especially on offshore rigs, where they are called simply day tour and night tour.

transmission *n*: the gear or chain arrangement by which power is transmitted from the prime mover to the drawworks, the mud pump, or the rotary table of a drilling rig.

trap *n*: a body of permeable oil-bearing rock surrounded or overlain by an impermeable barrier that prevents oil from escaping. The types of traps are structural, stratigraphic, or a combination of these. See *fault trap*.

traveling block *n*: an arrangement of pulleys, or sheaves, through which drilling line is strung and which moves up and down in the derrick or mast. See *block*.

treating *n*: to subject to a chemical or physical process or application to remove unwanted components.

trim *n*: the type of material used in wellhead control valves and chokes. For example, a valve made of stainless steel will have a stainless steel trim. *v*: to cut short pieces from an object.

trip *n*: the operation of hoisting the drill stem from and returning it to the wellbore. Shortened form of make a trip.

triple *n*: the modern term for a stand of three joints of the drill string left connected together. Also called thribble (obsolete).

tripping *n*: the operation of hoisting the drill stem out of and returning it to the wellbore. See *make a trip*.

tripping in *v*: to lower the drill stem, the tubing, or the casing into the wellbore.

tripping out *v*: to pull the drill stem out of the wellbore.

truncate *v*: to terminate abruptly.

tubing *n*: relatively small-diameter pipe that is run into a well to serve as a conduit for the passage of oil and gas to the surface.

tubing safety valve *n*: see *subsurface safety valve*.

tungsten carbide *n*: a fine, very hard, gray crystalline powder, a compound of tungsten and carbon. This compound is bonded with cobalt or nickel in cemented carbide compositions and used for cutting tools, abrasives, and dies.

tungsten carbide bit *n*: a type of roller cone bit with inserts made of tungsten carbide. Also called tungsten carbide insert bit.

tungsten carbide insert bit *n*: see *tungsten carbide bit*.

turnkey *adj*: the act of drilling a well under a turnkey drilling contract.

turnkey contract *n*: a drilling contract that calls for the payment of a stipulated amount to the drilling contractor on completion of the well. In a turnkey contract, the contractor furnishes all material and labor and controls the entire drilling operation, independent of operator supervision. A turnkey contract does not, as a rule, include the completion of a well as a producer.

turntable *n*: see *rotary table*.

U

uncased hole *n*: see *open hole*.

V

vacuum degasser *n*: a device in which gas is removed from the mud by the action of a vacuum inside a tank.

value moment *n*: more commonly called a safety moment or meeting. A meeting held before the start of a new job where the job process and duties of the people involved are reviewed. Any potential hazards are pointed out and remedies discussed. Additionally, the relevance or importance of the job to the outcome of the project is discussed.

V-door *n*: an opening at floor level of a derrick or mast shaped like an inverted V. The V-door is used as an entry to bring in drill pipe, casing, and other tools from the pipe rack.

voids *n pl*: cavities in a rock that do not contain solid material but may contain fluids. Also called vug.

vug *n*: 1. a cavity in a rock. 2. a small cavern, larger than a pore but too small to contain a person. Typically found in limestone subject to groundwater leaching.

W

waiting on cement (WOC) *adj*: pertaining to the time when drilling or completion operations are suspended so that the cement in a well can harden sufficiently.

walking beam *n*: the large steel or wooden beam that connects the power end of a cable-tool rig to the cable that suspended the bit in the well. The beam is suspended above the ground on a steel or wooden A-shaped frame called the Sampson Post. The walking beam is moved up and down by the power section of the rig which causes the bit on the cable to move up and down in the well striking the rock on the bottom in order to drill new hole.

wall cake *n*: a lining of clay particles that forms over the face of permeable formations that are drilled in a well if water-base drilling mud is being used. The clay lining restricts the loss of liquid from the drilling mud into the formation and acts as a stabilizing sheath over the formation.

water well *n*: a well drilled to obtain a fresh water supply to support drilling.

weight indicator *n*: an instrument near the driller's position on a drilling rig that shows both the weight of the drill stem that is hanging from the hook (hook load) and the weight that is placed on the bit by the drill collars (weight on bit).

well *n*: the hole made by the drilling bit, which can be open, cased, or both. Also called borehole, hole, or wellbore.

wellbore *n*: a borehole; the hole drilled by the bit. Also called a borehole or hole.

well completion *n*: 1. the methods of preparing a well for the production of oil and gas or for other purposes, such as injection; the method by which one or more flow paths for hydrocarbons are established between the reservoir and the surface. 2. the system of tubulars, packers, and other tools installed beneath the wellhead in the production casing—that is, the tool assembly that provides the hydrocarbon flow path or paths.

well control *n*: the methods used to control a kick and prevent a well from blowing out. Such techniques include, but are not limited to, keeping the borehole completely filled with drilling mud of the proper weight or density during all operations, exercising reasonable care when tripping pipe out of the hole to prevent swabbing, and keeping careful track of the amount of mud put into the hole to replace the volume of pipe removed from the hole during a trip.

wellhead *n*: the equipment installed at the surface of the wellbore.

well log *n*: see *log*.

well logging *n*: the recording of information about subsurface geologic formations, including records kept by the driller and records of mud and cutting analyses, core analysis, drill stem tests, and electric, acoustic, and nuclear procedures.

well site *n*: the place where a well is drilled.

well-sorted *adj*: grains of sand that are of roughly the same diameter.

whipstock *n*: a wedge that can be placed in a well to force the drill bit against the side of the hole so that the hole is deviated in a new direction.

whole core *n*: a large cylindrical rock sample that is retrieved by drilling or coring over the sample. The sample is extruded into a core barrel where it is retained while the drill string is pulled out of the hole to recover the core. The core is much larger in diameter and length than sidewall cores. Whole cores are more representative of the formation cored and are the preferred sample for detailed reservoir analysis.

whole coring *n*: see *whole core*.

wildcat *n*: a well drilled in an area where no oil or gas production exists.

wildcatter *n*: one who drills wildcat wells.

wireline *n*: 1. a cable capable of transmitting electric signals to tools suspended in the well by the cable. 2. a small-diameter metal (normally stainless steel) wire that does not transmit electric signals used to suspend lightweight tools in a well.

wireline operations *n pl*: the lowering of tools and gauges into the well for various purposes. Electric wireline operations, such as electric well logging and perforating, involve the use of electric wireline line or cable.

wire rope *n*: a rope made with twisted strands of wire. See *cable*.

WOC *abbr*: waiting on cement.

Index

accident statistics, 189

acetylene gas, 164

acidizing, 177

air hoists, 103

alternating current (AC) electricity, 98

American Petroleum Institute (API), 115

anchor points, 159

Anglo-Persian Oil Company, 13

annular preventer, 184

annular space, 126

annulus, 158

anticlinal traps, 62

API kelly, 115

Arctic sites, 82

Arctic submersibles, 24, 26–27

Arctic transport ships, 56

area drilling superintendent, 51

artificial lift hardware, 174

assistant drillers, 44

assistant rig superintendent, 44

automatic catheads, 102

automatic pipe racker, 152

backup tongs, 141

Baku, Azerbaijan, 7–8

barge control operators, 53

barge engineers, 53

barge masters, 53

barium sulfate(barite), 127

barrel, 10

bedrock, 9

beds, 60

bent housing, 180

Bissell, George, 8

bits

    about, 4–5

    diamond, 125

    drag, 122, 125

    matrix of, 125

    milling, 123

    polycrystalline diamond compact (PDC), 125

    roller cone, 122, 123–124

bit sub, 136

Black City, 7

blind rams, 184

blocks and drilling line, 104, 105, 108

blooie line, 128

blowout preventers (BOPs), 34, 159, 184

blowout preventers (BOPs) stack, 85

blowouts, 183, 184, 186

boilers, early day, 10

boreholes, 20

bottle-type submersibles, 25

bottomhole assembly (BHA), 120, 136

bottomhole location (BHL), 179

bottom plug, 158

bottom-supported rigs, 23

bottom-supported MODUs, 24

box-on-box substructure, 86

BP Thunder Horse rig, 54

brake bands, 5

breakout cathead, 104

breaking out pipe, 48

breakout tongs, 141

break tour, 92

British Petroleum, 13

Btus, 13

buck up (tighten), 143

bulk tanks, 133

bullwheel, 15

bumps, 159

bushing, 110-111

butane (C_4H_{10}), 56

cable-tool drillers, 11

cable-tool drilling, 15–16

applications for, 16

cable-tool rigs, 11

caisson, 26

California, 10

cased formations, 11

casing crews, 43, 154

casing pipe, 9

casing string, 162

catheads, 101, 102–103

catlines, 102

catshafts, 101

catwalk, 110

caving in, 9

cellars, 77

cementing casing, 43

cementing company, 43, 158

cementing head, 158

centralizers, 158

centrifuge, 132

characteristics of reservoir rocks, 58–59

choke manifold, 186

Christmas tree, 173

circulating equipment, 128–134

circulating system

circulating equipment, 128–134

drilling fluid, 126–128

clay, 11, 16, 127

coal vs. fuel oil, 10, 13

coiled tubing, 175

combination traps, 64

companies

drilling contractors, 39

drilling contracts, 39–40

operating companies, 38–39

service and supply companies, 40, 42–46

company representative, 51

completing a producing well, 173–174

completion rig, 174

conditioned mud, 20

conductor casing, 81

conductor holes, 80–81

conductor pipe, 81

confirmation wells, 69

connections, 121

core barrel, 170

coring, 170–172

crane operators, 52

crew safety, 50

crown block, 47

crude oil

about, 6

oil characteristics, 55, 57

origin of, 60

producers of, 38

cutters, 122

cuttings. See also fluid circulation; shale shakers

in early days, 11, 16

examination of, 163–172

treatment of, 74, 76, 130–131, 190

darcy, 59

Darcy, Henry, 59

daywork contracts, 40

deadline anchor, 104, 108

degasser, 132

density, 11

density log, 165

derrickman, 45–47, 49, 128, 133–134, 151, 160, 184

derrick raising, 89–91

derricks

    about, 2

    early day, 12, 15

    hoisting system, 109–110

    mast and derrick heights, 90–91

    standard, 89

desanders, 131

desilters, 131

development wells, 69

diamond bits, 125

direct current electricity, 98

directional drilling, 179–181

    rotary steerable assemblies, 181

    slide drilling with a motor, 180

directional holes, 118–119

doghouse, 91

dolomite, 59

doubles (pipeline), 89

downhole blowouts, 183

downhole motors, 18, 110, 118–119

drag bits, 122, 125

Drake, Edwin L., 9, 93

Drake Well, 9–10

drawworks, 88, 101–102

drawworks brake, 5

drawworks drum, 5

dressing off (activity), 182

drill ahead, 5

drill collars, 120

driller, 9

driller and assistant, 45

driller's controls, 45

drilling companies, 39

drilling contractors, 39

drilling contracts, 39–40

drilling crews, 44

drilling crew workshifts, 50

drilling engineer, 128

drilling fluid, 11, 20, 126–128

drilling line, 88

drilling mud, 12, 20

drilling operations, 135–162

    cementing, 158–159

    drilling ahead, 162

    running a surface casing, 154–158

    surface hole drilling, 135–147

    tripping in, 160–161

    tripping out with a kelly system, 148–151

    tripping out with a pipe racker, 152–154

    tripping out with a top-drive system, 152

drilling rate, 164

drilling rigs, 1

drill pipe, 4

drillships, 30, 33–36

drill site

    equipment moving, 82–84

    site preparation, 71–72, 73–81

drill stem testing, 168

drill stem testing (DST), 168

drill string, 5, 16, 120–122

drive pipe, 81

drums, 5

dynamic positioning operator, 34

dynamic positioning systems, 34

earthen pits, 73–76

electrical power transmission, 97–99

electric logs, 165

electric rig, 99

electric signals, 180

electro-magnetic (EM) signals, 180

elevator, 108

emergency medical technicians (EMTs), 52

environment

    concerns regarding, 189–190

    cuttings and, 130

    flaring and, 128

    mud impacts on, 127

    ocean, 39

    safeguards for, 73

    sensitive areas, 72

    specialists in, 72

    well site restoration, 173

equipment moving to drill site, 82–84

    land rigs, 82

    offshore rigs, 84

ethane (C_3H_6), 56
explorationists, 71
exploration wells, 69

fastline, 106
faults, 60
fault traps, 62
fingerboard, 151
fish, 181
fishermen, 181
fishing, 181–182
fishing jobs, 168
flapper, 156
flared gas, 128
float collara, 156
floaters, 33
floating units
    drill ships, 33–36
    semisubmersibles, 30–31
floes, 26
floorhands (rotary helpers/roughnecks), 44, 48–49
flow diverter, 136
fluid circulation, 19–20
footage rate contracts, 40
forging bit teeth, 123
formation, 10
formation evaluation
    coring, 170–172
    defined, 163
    drill stem testing, 168
    examining cuttings and drilling mud, 163–172
    well logging, 165–167
formation fluids, 20
foundation traps, 61
4-D seismic surveys, 68
fracturing, 177
friction catheads, 102
fuel oil vs. coal, 13
future of drilling, 191

gamma ray log, 165
gas
    about, 1
    natural gas, 6, 55
    origins and accumulation of, 60–61

propane (C_3H_8), 56
    in rocks, 164
gas characteristics
    liquefied natural gas (LNG), 56
    liquefied petroleum gas (LPG), 56
    natural gas, 55
    natural gas liquids (NGLs), 57
gas condensate, 57
gasoline, 57
gas seeps, 13
geologic forces, 60
geophones, 68
geophysicists, 68
Geronimo, 47
gooseneck, 116
grapples, 181
gravel packing, 178
guide shoe, 156
gushers, 12

heave compressors, 35
heavy lift vessels, 30
hexagonal kelly, 114
Higgins, Patillo, 11
history of drilling, 7–13
    California, 10
    Drake Well, 9–10
    Lucas Well, 11–12
    Spindletop well, 11–12, 20, 64–65
hoisting system, 100–110
    about, 100
    blocks and drilling line, 104, 105, 108
    catheads, 102–103
    drawworks, 101–102
    masts and derricks, 109–110
horizontal wells, 179
horsepower, 15
hot wire testing, 164
hydraulic fracturing, 177
hydraulic jars, 181
hydraulic motors, 18
hydrocarbons, 55
hydrochloric acid (HCL), 177
hydrocylones, 131
hydrofluoric acid (HFL), 177
hydrogas. See liquefied petroleum gas (LPG)

hydrogen sulfide, 55, 57
hydrophones, 68

independent companies, 38
infill wells, 69
inland barges, 27
intermediate casing, 162
International Association of Drilling
        Contractors (IADC), 40, 189
Iron Roughneck (commercial breakout
        machine), 147
island barge rigs, 27

jackups, 21, 24–26, 28–29, 92
jetted wastes, 74
joints, 4
joint setback, 153
junk, 181
junk mills, 182

kelly. See also rotating systems
        about, 15–16
        in drilling operation, 148–151
        hexagonal, 114
        limits with, 79
        and rotary table, 48, 110, 114–115
        square, 114
        storage of, 77
kelly drive bushing, 110–113
kelly is drilled down (condition), 139
kelly spinner, 142
kerogen, 60
kick off, 179, 184

lag testing, 164
land contractors, 39
land rigs, 22
        equipment moving to drill site, 82
leases, 38
limestone, 58
liners, 176
liquefied natural gas (LNG), 56
liquefied petroleum gas (LPG), 55, 56

log, 40
logging while drilling (LWD) tools, 167
Lucas, Anthony, 11
Lucas Well, 11–12

made up pipe, 79
major companies, 38
make a connection (activity), 139
make hole (activity), 93
makeup catheads, 104
make up pipe, 48
makeup tongs, 144
making hole, 5
marine crews, 52
marine risers, 34, 159
master bushing, 110
master bushing and kelly drive bushing, 111–113
mast load ratings, 91
masts, 2
        mast and derrick heights, 90–91
        raising, 89–91
matrix of bits, 125
measurement while drilling (MWD), 180
mechanical catheads, 102
mechanical power transmission, 97
mechanical rigs, 94
methane (CH$_3$), 55
metreage contracts, 40
milling bits, 123
mineral rights, 38
mobile offshore rigs, 23–36
mobile offshore rigs (MODUs)
        about, 23–24
        artic submersibles, 26–27
        bottle-type submersibles, 25
        bottom supported MODUs, 24
        floating units, 30–31
        island barge rigs, 27
        jackups, 28
        posted-barge submersibles, 24
        submersible MODUs, 24
monkeyboard, 46, 151
most hoisting system, 109–110
mouseholes, 79
mud
        about, 11

conditioned mud, 20
cuttings and drilling mud examination, 163–172
drilling mud, 12, 20
oil-base, 127
mud cleaning, 133
mud engineer, 47
mud log, 164
mud logger, 163
mud logging company, 40
mud motors, 118
mud pits, 184
mud pump, 97
mud tanks, 85
mud weight, 126
mule shoe, 181

natural gas, 6, 55
natural gas liquids (NGLs), 57
neutron log, 165
night toolpushers, 44
nippling BOPs, 159

off-bottom weight, 139
office personnel, 54
offshore drilling rigs, 2
offshore installation manager (OIM), 51
offshore personnel, 51–54
offshore rigging up activity, 92
offshore rigs, 84
oil, 1
    characteristics of, 55, 57
    origins and accumulation of, 60–61
    in rocks, 163
oil and gas reservoirs, 58–68
    characteristics of reservoir rocks, 58–59
oil-base mud, 127
oilwell drilling, early, 16
Oil Creek, 14
on-bottom weight, 139
open-hole completion, 176
operating companies, 38–39
operators, 38
origins and accumulation
    about, 60
    anticlinal traps, 62
    fault traps, 62
    finding petroleum traps, 65–68
    foundation traps, 61
    other traps, 64–65
    structural traps, 61
origins and accumulation of oil and gas, 60–61
outpost wells, 69
overshot, 181

packers, 168, 174
people, 44–54
    area drilling superintendent, 51
    company representative, 51
    crew safety, 50
    derrickman, 46–47, 49
    driller and assistant, 45
    drilling crews, 44
    drilling crew workshifts, 50
    floorhands (rotaryhelpers/roughnecks), 48–49
    office personnel, 54
    offshore personnel, 51–54
    rig superintendent and assistant, 44
perforating, 176–177
perforating gun, 176–177
perforations, 176
permafrost, 73
permeable rocks, 59
personal protective gear (PPE), 3, 50
petrochemical feedstock, 57
petroleum era, 10
petroleum geology, 65
petroleum traps discovery, 65–68
piercement salt dome, 64
pile driver, 81
pins, 112
pipe rams, 184
pipe tally, 136
plastic material, 65
platform, 21
plugging and abandoning wells (P&A), 173
plug stops, 158
polycrystalline diamond compact (PDC) bits, 125
pontoons, 30
porosity, 58
porous rock, 58
posted-barge submersibles, 24

power rigs, 94

power swivel, 117

power system

about, 93–99

electrical power transmission, 97–99

mechanical power transmission, 97

prime movers, 95

production, 10

production casing, 162

production tubing, 173, 174–175

production wellhead, 35

propane (C_3H_8), 56

proppants, 177

pulleys, 106

pull line, 141

pumps, 19

quadruples (pipeline), 89

racked back stand, 150

rate of penetration (ROP), 123

rathole rig, 78

ratholes, 77–78

recycling, 76

refined hydrocarbon, 57

remotely operated vehicle (ROV), 35

repeat formation tester (RFT), 168

reserve pits, 74

reservoir stimulating, 177

reservoir tapping, 1

reservoir treating, 177

rig components, 93–99

circulating system, 126–134

hoisting system, 100–110

power system, 93–99

rotating systems, 110–126

rig floor, 4

rigged-down status, 82

rigged up status, 78

rigged-up status, 82

rigging up activity, 85–92

additional equipment, 91–92

mast or derrick raising, 89–91

offshore, 92

substructures, 85–88

rig manager, 3, 44

rig safety and environmental concerns, 189–190

rig superintendent, 3

rig superintendent and assistant, 44

rig supervisors, 44

rocks

gas in, 164

oil in, 163

roller cone bits, 122, 123–124

rotary drilling rig, 11, 16–20

fluid circulation, 19–20

rotating systems, 18–19

rotary drilling rigs, types of

land rigs, 22

mobile offshore rigs, 23–36

rotary helpers, 44, 48

rotary speed, 179

rotary steerable assemblies, 179, 181

rotary table, 18

rotary-table systems, 110, 152

rotating components, 97

rotating head, 128

rotating systems, 18–19, 110–126

bits, 122

downhole motors, 118–119

drag bits, 125

drill string, 120–122

kelly, 114–115

master bushing and kelly drive

bushing, 111–113

roller cone bits, 123–124

rotary-table systems, 110

swivel, 115–117

top drives, 115–118

turntables, 111

weight on bits and rotating speeds, 126

rotors, 118

roughnecks, 44

round trip, 135

roustabouts, 52

royalties, 39

run in activity, 89

running quicksand, 11

run pipe, 111

safety equipment, 47

safety training frequency, 189

salt, 65

salt domes, 65

sand reel, 16

sandstone, 58

SCR house, 98

sediment, 57

seismic signals, 68

seismic surveys, 68

seismology, 68

self-propelled barges, 27

semisubmersibles, 30–31

Seneca Oil Company, 9

service and supply companies, 40, 42–46

shakers, 131

shale, 65

shale diapirs, 65

shale pits, 74

shale shakers, 128

shaped charges, 176

shear rams, 184

sheaves, 106

shut in well, 136

sidewall coring, 170

silicon controlled rectification (SCR) rigs, 94

singles (pipeline), 89

sinker bar, 15

site preparation, 73–81
    cellars, 77
    conductor holes, 80–81
    earthen pits, 73–76
    mouseholes, 79
    ratholes, 77–78
    surface preparation, 73

site selection, 71–72

skidding of rigs, 21

skid the rig step, 82

slack off, 5

slide drilling with a motor, 179, 180

slingshot substructure, 86

slurry, 158

Smith, William A., 9

soft elevating-substructure, 86

sonic logs, 165

sour crude, 57

sour gas, 55

spears, 182

special operations, 179–188
    directional drilling, 179–181
    fishing, 181–182
    well control, 183–188

Spindletop well, 11–12, 20, 64–65

spool, 88

spudded wells, 11

square kelly, 114

stabbing the drill string, 117, 156

standard derricks, 89

standpipe, 130

stators, 118

steering tools, 180

stem, 115

step-out wells, 69

structural traps, 61

submersible MODUs, 24

types of, 24

subsea blowout preventers (BOPs), 52, 186

subsea engineers, 52

subsea equipment supervisors, 52

substructures, 4, 85–88

subsurface safety valve (SSSV), 175

sulfides, 57

supply reel, 104

surface blowout, 183

surface casing, 136

surface hole, 136

surface location (SL), 179

surface preparation, 73

swamp barges, 27

swampers, 91

sweet crude, 57

swivel, 110, 115–117

swivel stem, 116

tapered bowl, 112

tectonic plates, 60

tensile forces, 121

tensioners, 35

threads, 121

thrusters, 32

thumpers, 68

Tinkerbell line, 47

Titusville, PA, 14

tongs, 48

tons, 2

tool joint, 108

toolpush, 3

toolpushers, 3, 44

top drives, 18, 117–118

    advantages of, 118

top-drive systems, 110

top drive system vs. rotary-table system, 152

top plugs, 159

topside, 92

torque, 99

total depth (TD), 162

tours, 49

Townsend, James M., 9

transmission lines, 99

traps, 60, 64–65

traveling blocks, 100

trim, 54

triples (pipeline), 89

tripping in, 46

tripping out, 46

trips, 108

truncated layers, 62

turnkey basis contracts, 40

turntables, 18, 111

vacuum degasser, 132

value moment, 50

V-door, 110

vugs, 59

waiting on cement (WOC) condition, 159

walking bean, 15

wall cake, 12

water wells, 75

weight indicator, 139

weight on bits and rotating speeds, 126

wellbore, 111

well completion, 136, 173–178

    completing a producing well, 173–174

    perforating, 176–177

    plugging and abandoning wells, 173

    production tubing, 174–175

    well testing and treating, 177–178

well control, 183–188

wellhead equipment, 128

well logging, 165–167

well logging company, 42

wells, 1

well site, 9

well-sorted rock grains, 58

well testing and treating

    acidizing, 177

    fracturing, 177

    gravel packing, 178

well types, 69

whipstocks, 179

white gas, 57

whole coring, 170

wildcatters, 65

wild cat wells, 65

wireline, 166

work shifts, 50